The Physics of Laser Plasma Interactions

CRC Press
Taylor & Francis Group
Boca Raton London New York

CRC Press is an imprint of the
Taylor & Francis Group, an **informa** business

First published 2003 by Westview Press

Published 2018 by CRC Press
Taylor & Francis Group
6000 Broken Sound Parkway NW, Suite 300
Boca Raton, FL 33487-2742

CRC Press is an imprint of Taylor & Francis Group, an Informa business

No claim to original U.S. Government works

A Cataloging-in-Publication data record for this book is available from the Library of Congress.

ISBN 13: 978-0-8133-4083-8 (pbk)

Visit the Taylor & Francis Web site at
http://www.taylorandfrancis.com

and the CRC Press Web site at
http://www.crcpress.com

Frontiers in Physics
David Pines, Editor

Volumes of the Series published from 1961 to 1973 are not officially numbered. The parenthetical numbers shown are designed to aid librarians and bibliographers to check the completeness of their holdings.

Titles published in this series prior to 1987 appear under either the W. A. Benjamin or the Benjamin/Cummings imprint; titles published since 1986 appear under the Westview Press imprint.

Volumes published from 1974 onward are being numbered as an integral part of the bibliography.

Frontiers in Physics

Frontiers in Physics

Editor's Foreword

The problem of communicating in a coherent fashion recent developments in the most exciting and active fields of physics continues to be with us. The enormous growth in the number of physicists has tended to make the familiar channels of communication considerably less effective. It has become increasingly difficult for experts in a given field to keep up with the current literature; the novice can only be confused. What is needed is both a consistent account of a field and the presentation of a definite "point of view" concerning it. Formal monographs cannot meet such a need in a rapidly developing field, while the review article seems to have fallen into disfavor. Indeed, it would seem that the people most actively engaged in developing a given field are the people least likely to write at length about it.

FRONTIERS IN PHYSICS was conceived in 1961 in an effort to improve the situation in several ways. Leading physicists frequently give a series of lectures, a graduate seminar, or a graduate course in their special fields of interest. Such lectures serve to summarize the present status of a rapidly developing field and may well constitute the only coherent account available at the time. Often, notes on lectures exist (prepared by the lecturer himself, by graduate students, or by postdoctoral fellows) and are distributed in mimeographed form on a limited basis. One of the principal purposes of the FRONTIERS IN PHYSICS Series is to make such notes available to a wider audience of physicists.

It should be emphasized that lecture notes are necessarily rough and informal, both in style and content; and those in the Series will prove no exception. This is as it should be. One point of the Series is to offer new, rapid, more informal, and, it is hoped, more effective ways for physicists to teach one another. The point is lost if only elegant notes qualify.

The above words, written some twenty-five years ago, continue to be applicable. During this period the field of laser plasma interactions has emerged as a major sub-field of applied plasma physics. William Kruer has been a leading contributor to our understanding of laser plasma physics. He is thus especially well qualified to provide an introductory overview of this important and active field. His book contains a clear, physically motivated, description of the major physical processes which determine the interaction of intense lightwaves with plasmas. It should thus prove useful not only as a text for an introductory graduate course in plasma physics, but as a reference book for all scientists interested in plasma phenomena.

David Pines
Urbana, Illinois
August 1987

Preface

The subject of this book is the physics of laser plasma interactions. This exciting field of applied plasma physics has received great stimulation from nearly two decades of research in laser fusion. The field has been a fruitful test bed for exploring a wide range of basic plasma phenomena, including the excitation of plasma waves, generation and saturation of plasma instabilities, and transport of intense heat fluxes. Numerous nonlinear effects have been characterized and observed in experiments, and regimes of optimum coupling have been identified and tested. Hence this book is of interest not only to those active in the use of high power lasers, but also to scientists in many other fields where plasma phenomena are involved.

My aim has been to give a clear, physically-motivated treatment of the major processes. Since the subject is of interest to scientists with many different specialties, very little prior knowledge of plasma physics is assumed. In Chapter 1, basic plasma concepts are introduced and a theoretical description of plasmas is developed. In Chapter 2, a complementary and very useful numerical model of plasmas is presented. These two complementary levels of description are then used to describe laser plasma interactions. Chapters 3, 4, and 5 treat the linear theory of light wave propagation in plasmas, including linear mode conversion into plasma waves and collisional damping. The excitation of a variety of plasma instabilities by intense light waves is then treated in Chapters 6, 7, and 8. In Chapters 9, 10, and 11, important nonlinear consequences of the various processes are discussed using both simple theoretical models and computer simulations. The physics of electron heat transport in laser-produced plasma is discussed in Chapter 12. Finally, some experimental observations of the various laser plasma processes are discussed in Chapter 13.

This manuscript is based on lectures given in a graduate course in the Department of Applied Science of the University of California, Davis. A detailed review of laser plasma interactions is beyond the scope of this book. However, a broad cross-section of references to the literature is given, particularly in those areas of very active research. Lastly, I do not consider either implosion physics or the very rich topic of electromagnetic waves in magnetized plasmas.

I am grateful to numerous colleagues with whom I have worked, and especially to present and former members of the plasma physics group in the laser fusion program at the Lawrence Livermore National Laboratory: J. Albritton, J. Denavit, K. Estabrook, R. Faehl, D. Hewett, A. B. Langdon, B. Lasinski, W. Mead, C. Max, C. Randall, J. Thomson, E. Valeo and E. Williams. I also thank many others for helpful comments on portions of the manuscript, including J. M. Dawson, E. M. Campbell, H. Baldis, R. P. Drake, T. L. Crystal, C. S. Liu, R. Turner, D. Phillion, M. Rosen, J. DeGroot, J. Delettrez, L. Goldman, T. Tajima, and R. Lehmberg. I acknowledge the encouragement of J. Nuckolls and J. Lindl. S. Auguadro typed the original lecture notes. T. L. Crystal very ably produced the final manuscript. He and A. Wylde provided just the right amount of support and pressure to finish. I am grateful to the Lawrence Livermore National Laboratory, and particularly to B. Quick and P. Brown for a variety of assistance. Finally, I warmly thank my family for generously allowing me to devote many evenings and weekends to this manuscript.

Contents

CHAPTER 1

Basic Concepts and Two-Fluid Description of Plasmas

The study of the interaction of intense laser light with plasmas serves as an excellent introduction to the field of plasma physics. Both the linear and nonlinear theory of plasma waves, instabilities and wave-particle interactions are important for understanding the laser plasma coupling. Indeed, the field is a veritable testing ground for many fundamental processes. Numerous plasma effects have now been observed in laser plasma experiments, and many challenging problems remain to be understood.

Since laser plasma interactions are of interest to scientists from many different fields of expertise, little prior background in plasma physics will be assumed. Even for those with plasma experience, it can be very instructive and refreshing to begin from the basics and examine a field of applications. Two levels of description will be used – a theoretical one based on the two-fluid theory of plasmas and a numerical one based on particle simulation codes. These two descriptions both reinforce and complement one another. For example, the particle simulations allow one to both test the theory and develop some understanding of the nonlinear effects.

1.1 BASIC PLASMA CONCEPTS

Let's begin. A plasma is basically just a system of N charges which are coupled to one another via their self-consistent electric and magnetic fields. Consider then following the evolution of these N charges. Even neglecting magnetic fields and electromagnetic waves, we must in principle solve 6N coupled equations:

$$m_i \ddot{\mathbf{r}}_i = q_i \mathbf{E}(\mathbf{r}_i)$$

$$\mathbf{E}(\mathbf{r}_i) = \sum_j \frac{q_j}{|\mathbf{r}_i - \mathbf{r}_j|^3}(\mathbf{r}_i - \mathbf{r}_j).$$

Here m_i, q_i and $\mathbf{r}_i$ are the mass, charge and position of the ith particle, and $\mathbf{E}$ is the electrostatic field. This is clearly an unpromising approach if a nontrivial number of charges is considered.

Fortunately a very great simplification is possible if we focus our attention on collisionless plasma behavior. We can decompose the electric field into two fields ($\mathbf{E}_1$ and $\mathbf{E}_2$) which have distinct spatial scales. The field $\mathbf{E}_1$ has spatial variations on a scale length much less than the so-called electron Debye length, which is the length over which the field of an individual charge is shielded out by the response of the surrounding charges. $\mathbf{E}_1$ represents the rapidly fluctuating microfield due to multiple and random encounters (collisions) among the discrete charges. In contrast, $\mathbf{E}_2$ represents the field due to deviations from charge neutrality over space scales greater than or comparable to the Debye length. This field gives rise to "collective" or coherent motion of the charges.

We thus have a natural separation into collisional and collective behavior. Not surprisingly, the collisional behavior becomes negligible when the number of electrons in a sphere with a radius equal to the electron Debye length becomes very large. To motivate this, let us carry out a simple calculation of electron scattering by ions. As illustrated in Fig. 1.1, we consider an electron with velocity v, mass m and charge e streaming past an ion with charge Ze. The distance of closest approach is b. The electron undergoes a change in velocity Δv which is approximately

$$\Delta v = \frac{Z e^2}{m b^2}\left(\frac{2b}{v}\right),$$

which is just the maximum electrostatic force times the interaction time ($\sim 2b/v$). If we assume many randomly spaced ions, $\langle \Delta v \rangle = 0$, where the brackets denote an average. However, there is a change in the mean square velocity. This average rate of change is given by $(\Delta v)^2$ times the rate of encounters, which is $n_i \, \sigma \, v$. Here n_i is the ion density and σ is the cross-section of impact. Summing over all encounters gives

$$\frac{\mathrm{d}}{\mathrm{d}\,t} \langle \, (\Delta v)^2 \, \rangle \;=\; \int \, 2\pi \, b \; \mathrm{d}b \; n_i \, v \, (\Delta v)^2 \; .$$

If we substitute for Δv and integrate over impact parameters, we obtain

$$\langle \, (\Delta v)^2 \, \rangle \;=\; \frac{8\pi \, n_i \, Z^2 e^4 \, \ln \Lambda}{m^2 v} \, t \, ,$$

where Λ is the ratio of the maximum and minimum impact parameters (b_{max} and b_{min}). The maximum impact parameter is approximately the electron Debye length, since other electrons in the plasma shield out the Coulomb potential over this distance. The minimum impact parameter is the larger of either the classical distance of closest approach ($b_{\mathrm{min}} \approx Ze^2/mv^2$) or the DeBroglie wavelength of the electron ($b_{\mathrm{min}} \approx \hbar/mv$), where $\hbar$ is Planck's constant. Using the first, the distance of closest approach, we have $\Lambda \approx 9 N_D/Z$, where N_D is the number of electrons in a Debye sphere. In particular $N_D = \frac{4}{3}\pi \, n_e \lambda_{\mathrm{De}}^3$, where n_e is the electron density and λ_{De} is the electron Debye length. This important length will be derived later in this chapter.

It is convenient to define a ninety-degree deflection time (t_{90°) by the condition that the root-mean-square change in velocity becomes as large as the velocity. Hence

$$t_{90^\circ} \;=\; \frac{m^2 v^3}{8\pi \, n_i \, Z^2 e^4 \, \ln \Lambda} \; .$$

Averaging over a Maxwellian distribution of velocities then provides us with a convenient measure of the mean rate ($\nu_{90^\circ} \equiv 1/t_{90^\circ}$) at which electron-ion collisions scatter electrons through a large angle:

$$\nu_{90^\circ} \;=\; \frac{8\pi \, n_i \, Z^2 e^4 \, \ln \Lambda}{6.4 \, m^2 v_e^3} \; . \tag{1.1}$$

Here $v_e = \sqrt{\theta_e/m}$ is the electron thermal velocity and θ_e is the electron

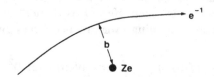

Figure 1.1   An electron is deflected as it streams past an ion.

temperature. We note that

$$\frac{\nu_{90°}}{\omega_{\text{pe}}} \simeq \frac{Z \ln \Lambda}{10} \frac{1}{N_D},$$

where ω_{pe} is the electron plasma frequency, which we will see is a frequency characteristic of collective electron motion.

The important point we wish to make is now apparent. The fine scale, collisional interactions can be neglected to zeroth order in the parameter $1/N_D$. If we express the electron density in cm^{-3} and the electron temperature in eV, then $N_D = 1.7 \times 10^9 \, (\theta_e^3/n_e)^{1/2}$. N_D can be very large even in a rather dense plasma, provided the electron temperature is high. For example, if $n_e = 10^{21}$ cm^{-3} and $\theta_e = 1$ keV, $N_D \approx 1700$. In the collisionless limit ($N_D \to \infty$), the fine scale fluctuating microfields associated with discrete charges are completely negligible. The plasma behavior can then be investigated by solving for the motion of the charges in the smoothed or coarse-grained fields which arise from the collective motion of large numbers of charges.

We will develop two parallel levels of description for the collective behavior. One level is analytical. Starting from the Vlasov equation, we will derive moment (fluid-like) equations for the electrons and ions by averaging over the velocities of the charges. This so-called two-fluid description will then be used extensively to describe a wide variety of laser plasma interactions. The second level of description is numerical: the use of particle simulations. These simulations are a powerful tool for investigating nonlinear effects and kinetic effects (effects which depend on the details of the velocity distribution of the particles).

1.2   THE VLASOV EQUATION

The natural starting point for describing the evolution of a collisionless plasma is the Vlasov equation. We first introduce the phase space distribution function $f_j(\mathbf{x}, \mathbf{v}, t)$. This is simply the function which characterizes the location of the particles of species j in phase space $(\mathbf{x}, \mathbf{v})$ as a function of time. Knowing the laws of motion, we can readily derive an equation for $f_j(\mathbf{x}, \mathbf{v}, t)$. Since particles are assumed to be neither created nor destroyed as they move from one location in phase space to another (no ionization or recombination), $f_j(\mathbf{x}, \mathbf{v}, t)$ must obey the continuity equation:

$$\frac{\partial f_j}{\partial t} + \frac{\partial}{\partial \mathbf{x}} \cdot (\dot{\mathbf{x}} f_j) + \frac{\partial}{\partial \mathbf{v}} \cdot (\dot{\mathbf{v}} f_j) = 0 \,. \tag{1.2}$$

From the laws of motions, we have

$$\dot{\mathbf{x}} = \mathbf{v}$$
$$\dot{\mathbf{v}} = \frac{q_j}{m_j} \left(\mathbf{E} + \frac{\mathbf{v} \times \mathbf{B}}{c} \right) \,, \tag{1.3}$$

where q_j and m_j are the charge and mass of the jth species and $\mathbf{E}$ and $\mathbf{B}$ are the coarse-grained fields associated with the collective behavior. Noting that $\mathbf{x}$ and $\mathbf{v}$ are independent variables and substituting Eq. (1.3) into Eq. (1.2), we arrive at the Vlasov equation:

$$\frac{\partial f_j}{\partial t} + \mathbf{v} \cdot \frac{\partial f_j}{\partial \mathbf{x}} \cdot + \frac{q_j}{m_j} \left(\mathbf{E} + \frac{\mathbf{v} \times \mathbf{B}}{c} \right) \cdot \frac{\partial f_j}{\partial \mathbf{v}} = 0 \,. \tag{1.4}$$

This equation simply says that $f_j(\mathbf{x}(t), \mathbf{v}(t), t)$ is a constant; i.e., the phase space density is conserved following a dynamical trajectory. Such an equation applies to each charge species in the plasma.

The Vlasov equation, augmented with Maxwell's equations, is a complete description of collisionless plasma behavior. In practice, we need a more tractable description which can be obtained by averaging over the velocities of the individual particles. By taking different velocity moments of the Vlasov equation, we can derive equations for the evolution in space and time of the density, mean velocity, and pressure of each species. As we will see, each moment brings in the next higher moment, generating an infinite set of moment equations. However, we can fortunately truncate the series of equations by introducing assumptions about the heat flow.

1.3   THE MOMENT EQUATIONS

Let us now derive the moment equations and motivate their truncation. First, we note that the density (n_j), mean velocity (u_j), and pressure tensor $(\underset{\sim}{P}_j)$ are determined by averaging the various moments of the phase space distribution function over velocities:

$$n_j = \int f_j(\mathbf{x}, \mathbf{v}, t) \, d\mathbf{v} \qquad (1.5)$$

$$n_j \, \mathbf{u}_j = \int \mathbf{v} \, f_j(\mathbf{x}, \mathbf{v}, t) \, d\mathbf{v} \qquad (1.6)$$

$$\underset{\sim}{P}_j = m_j \int (\mathbf{v} - \mathbf{u}_j)(\mathbf{v} - \mathbf{u}_j) \, f_j(\mathbf{x}, \mathbf{v}, t) \, d\mathbf{v} \, . \qquad (1.7)$$

In deriving the moment equations, we will suppress the subscript j, since it is clear that these equations will apply to each charge species. Averaging the Vlasov equation over velocity gives

$$\int d\mathbf{v} \left[\frac{\partial f}{\partial t} + \mathbf{v} \cdot \frac{\partial f}{\partial \mathbf{x}} + \frac{q}{m} \left(\mathbf{E} + \frac{\mathbf{v} \times \mathbf{B}}{c} \right) \cdot \frac{\partial f}{\partial \mathbf{v}} \right] = 0 \, . \qquad (1.8)$$

The first two terms in Eq. (1.8) give

$$\int d\mathbf{v} \, \frac{\partial f}{\partial t} = \frac{\partial n}{\partial t}$$

$$\int d\mathbf{v} \, \mathbf{v} \cdot \frac{\partial f}{\partial \mathbf{x}} = \int d\mathbf{v} \sum_i v_i \frac{\partial f}{\partial x_i}$$

$$= \frac{\partial}{\partial \mathbf{x}} \cdot (n \, \mathbf{u}) \, .$$

The third term in Eq. (1.8) vanishes, as can be seen by integrating by parts and noting that $f \to 0$ as $|\mathbf{v}| \to \infty$. Hence the first moment of the Vlasov equation gives the continuity equation for the particle density:

$$\frac{\partial n}{\partial t} + \frac{\partial}{\partial \mathbf{x}} \cdot (n \, \mathbf{u}) = 0 \, . \qquad (1.9)$$

The next moment of the Vlasov equation is

$$\int d\mathbf{v} \, \mathbf{v} \left[\frac{\partial f}{\partial t} + \mathbf{v} \cdot \frac{\partial f}{\partial \mathbf{x}} + \frac{q}{m} \left(\mathbf{E} + \frac{\mathbf{v} \times \mathbf{B}}{c} \right) \cdot \frac{\partial f}{\partial \mathbf{v}} \right] = 0 \, . \qquad (1.10)$$

The first term in Eq. (1.10) is straightforward:

$$\int d\mathbf{v}\, \mathbf{v}\, \frac{\partial f}{\partial t} = \frac{\partial}{\partial t} n\mathbf{u}\,.$$

The second term gives

$$\int d\mathbf{v}\, \mathbf{v}\,\mathbf{v}\cdot \frac{\partial f}{\partial \mathbf{x}} = \frac{\partial}{\partial \mathbf{x}}\cdot \int d\mathbf{v}\, \mathbf{v}\,\mathbf{v}\, f$$

$$= \frac{\partial}{\partial \mathbf{x}}\cdot \left(\frac{\underset{\approx}{\mathbf{P}}}{m} + n\,\mathbf{u}\,\mathbf{u}\right).$$

This result is readily obtained by rewriting the integral as

$$\int d\mathbf{v}\, (\mathbf{v} - \mathbf{u} + \mathbf{u})(\mathbf{v} - \mathbf{u} + \mathbf{u})\, f = \frac{\underset{\approx}{\mathbf{P}}}{m} + n\,\mathbf{u}\,\mathbf{u}\,,$$

since $\int (\mathbf{v} - \mathbf{u})\, f\, d\mathbf{v} = 0$. Evaluation of the last term in Eq. (1.10) yields

$$\int d\mathbf{v}\, \mathbf{v}\, \frac{q}{m}\left(\mathbf{E} + \frac{\mathbf{v}\times\mathbf{B}}{c}\right)\cdot \frac{\partial f}{\partial \mathbf{v}} = -\frac{n\,q}{m}\left(\mathbf{E} + \frac{\mathbf{u}\times\mathbf{B}}{c}\right),$$

where we have integrated by parts. Collecting the above terms, we obtain the equation of motion for the charged fluid:

$$\frac{\partial}{\partial t}(n\,\mathbf{u}) + \frac{\partial}{\partial \mathbf{x}}\cdot(n\,\mathbf{u}\,\mathbf{u}) = \frac{n\,q}{m}\left(\mathbf{E} + \frac{\mathbf{u}\times\mathbf{B}}{c}\right) - \frac{\partial}{\partial \mathbf{x}}\cdot \frac{\underset{\approx}{\mathbf{P}}}{m}\,. \qquad (1.11)$$

It is convenient to rewrite the first two terms of Eq. (1.11) using the continuity equation and to assume that the pressure is isotropic, i.e., $\underset{\approx}{\mathbf{P}} = \underset{\approx}{\mathbf{I}}\, p$ where $\underset{\approx}{\mathbf{I}}$ is the unit dyad. Then

$$n\,\frac{\partial \mathbf{u}}{\partial t} + n\,\mathbf{u}\cdot \frac{\partial \mathbf{u}}{\partial \mathbf{x}} = \frac{n\,q}{m}\left(\mathbf{E} + \frac{\mathbf{u}\times\mathbf{B}}{c}\right) - \frac{1}{m}\frac{\partial p}{\partial \mathbf{x}}\,. \qquad (1.12)$$

Observe that each moment brings in the next higher one. The continuity equation for the density involves the mean velocity; the force equation for the velocity brings in the pressure. The next moment will give us an equation for the pressure (energy density) which involves the heat flow. Continuing, we would end up with an infinite set of coupled equations, hardly a practical description.

Fortunately, we can truncate the moment equations by making various assumptions about the heat flow, which gives us a so-called equation of state. The simplest assumption is that the heat flow is so rapid that the temperature of the charged fluid is a constant. In this case, we have the isothermal equation of state: $p = n\theta$, where the temperature θ is a constant. This equation of state, plus the continuity and force equations for the fluid, and Maxwell's equations form a closed description.

The isothermal equation of state is appropriate when $\omega/k \ll v_t$, where ω and k are the frequency and wave number characteristic of the physical process being considered and v_t is the thermal velocity of the particles. In the opposite limit $(\omega/k \gg v_t)$ we can simply neglect the heat flow. This assumption leads to an adiabatic equation of state, which we will now derive.

To obtain an equation for the pressure, we multiply the Vlasov equation by the kinetic energy and average over velocity:

$$\int \frac{mv^2}{2} \, d\mathbf{v} \left[\frac{\partial f}{\partial t} + \mathbf{v} \cdot \frac{\partial f}{\partial \mathbf{x}} + \frac{q}{m} \left(\mathbf{E} + \frac{\mathbf{v} \times \mathbf{B}}{c} \right) \cdot \frac{\partial f}{\partial \mathbf{v}} \right] = 0 \,. \quad (1.13)$$

At this point, let us specialize to one-dimension to simplify the algebra. The first term can be written as

$$\frac{m}{2} \frac{\partial}{\partial t} \int f \, (v - u + u)^2 \, dv = \frac{1}{2} \frac{\partial}{\partial t} (p + nmu^2) \,.$$

The next term in Eq. (1.13) gives

$$\frac{m}{2} \frac{\partial}{\partial x} \int f \, (v - u + u)^3 \, dv = \frac{\partial Q}{\partial x} + \frac{3}{2} \frac{\partial}{\partial x} (up) + \frac{m}{2} \frac{\partial}{\partial x} (nu^3) \,,$$

where $Q \equiv (m/2) \int (v - u)^3 f \, dv$. The final term in Eq. (1.13) is simply

$$\frac{q}{2} \int v^2 E \frac{\partial f}{\partial v} \, dv = -nquE \,.$$

Collecting terms, we obtain

$$\frac{1}{2} \frac{\partial}{\partial t} (p + nmu^2) + \frac{3}{2} \frac{\partial}{\partial x} (up) + \frac{1}{2} \frac{\partial}{\partial x} (nmu^3) + \frac{\partial Q}{\partial x} = qnuE \,. \quad (1.14)$$

A great deal of simplification results from the use of the lower moment equations. In particular,

$$\frac{\partial}{\partial t} \left(\frac{n\,m\,u^2}{2} \right) = \frac{m\,u^2}{2} \frac{\partial n}{\partial t} + n\,m\,u \frac{\partial u}{\partial t}\,.$$

Using Eqs. (1.9) and (1.12), substituting into Eq. (1.14), and cancelling terms gives

$$\frac{\partial p}{\partial t} + u\frac{\partial p}{\partial x} + 3p\frac{\partial u}{\partial x} + 2\frac{\partial Q}{\partial x} = 0\,. \tag{1.15}$$

To obtain the adiabatic equation of state, we neglect the heat flow. This assumes that $\partial Q/\partial x$ is much less than the other terms in Eq. (1.15). For example, demanding that $\partial Q/\partial x \ll \partial p/\partial t$ gives $\omega p \gg k Q$, where ω and k are a frequency and wavenumber characteristic of the process being considered. Clearly $Q < Q_{\max} \sim n\,\theta\,v_t$, where v_t is the thermal velocity. Hence, to neglect heat flow it is sufficient to assume that $\omega/k \gg v_t$.

With this assumption, Eq. (1.15) reduces to

$$\frac{\partial p}{\partial t} + u\frac{\partial p}{\partial x} + 3p\frac{\partial u}{\partial x} = 0\,. \tag{1.16}$$

The continuity equation allows us to express $\partial u/\partial x$ as

$$\frac{\partial u}{\partial x} = -\left(\frac{\partial}{\partial t} + u\frac{\partial}{\partial x} \right) \ln\,n\,. \tag{1.17}$$

Substituting Eq. (1.17) into Eq. (1.16) gives

$$\left(\frac{\partial}{\partial t} + u\frac{\partial}{\partial x} \right) \ln\,p - \left(\frac{\partial}{\partial t} + u\frac{\partial}{\partial x} \right) \ln\,n^3 = 0\,,$$

or

$$\left(\frac{\partial}{\partial t} + u\frac{\partial}{\partial x} \right) \frac{p}{n^3} = 0\,. \tag{1.18}$$

This equation shows that, following the plasma flow, $p/n^3 = $ constant, which is the adiabatic equation of state for motion with one degree of freedom. This equation of state is readily generalized to $p/n^\gamma = $ constant, where $\gamma = (2+N)/N$ and N is the number of degrees of freedom.

1.4 THE TWO-FLUID DESCRIPTION OF PLASMA

Finally let us summarize the fluid equations, which we have derived by taking moments with respect to velocity of the Vlasov equation. The first two equations are the continuity and force equations for the density and mean velocity of particles with charge q_j and mass m_j.

$$\frac{\partial n_j}{\partial t} + \frac{\partial}{\partial \mathbf{x}} \cdot (n_j \, \mathbf{u}_j) = 0 \qquad (1.19)$$

$$n_j \left(\frac{\partial \mathbf{u}_j}{\partial t} + \mathbf{u}_j \cdot \frac{\partial \mathbf{u}_j}{\partial \mathbf{x}} \right) = \frac{n_j \, q_j}{m_j} \left(\mathbf{E} + \frac{\mathbf{u}_j \times \mathbf{B}}{c} \right) - \frac{1}{m_j} \frac{\partial p_j}{\partial \mathbf{x}} . \qquad (1.20)$$

The pressure of each charged fluid is related to its density by an equation of state, which depends on the characteristic frequency (ω) and wavenumber (k) of the process being considered. When $\omega/k \ll v_j$, the isothermal equation of state is valid:

$$p_j = n_j \, \theta_j , \qquad (1.21)$$

where θ_j is the constant value of the temperature and $v_j = \sqrt{\theta_j/m_j}$. When $\omega/k \gg v_j$, the adiabatic equation of state obtains:

$$\frac{p_j}{n_j^\gamma} = \text{constant} , \qquad (1.22)$$

where $\gamma = (2 + N)/N$ and N is the number of degrees of freedom. When $\omega/k \sim v_j$, the details of the velocity distribution of the charges are important. The fluid description is then inadequate, and we must return to the Vlasov equation.

For a plasma composed of electrons and one species of ions, Eqs. (1.19-1.22) constitute the well-known two-fluid model. This description is completed by Maxwell's equations, which relate the electric and magnetic fields to the charge and current densities of the plasma. In cgs units, Maxwell's equations are

$$\nabla \cdot \mathbf{E} = 4 \pi \rho \qquad (1.23)$$

$$\nabla \cdot \mathbf{B} = 0 \qquad (1.24)$$

$$\nabla \times \mathbf{E} = -\frac{1}{c} \frac{\partial \mathbf{B}}{\partial t} \qquad (1.25)$$

$$\nabla \times \mathbf{B} = \frac{4 \pi}{c} \mathbf{J} + \frac{1}{c} \frac{\partial \mathbf{E}}{\partial t} , \qquad (1.26)$$

where $\rho = \sum_j n_j \, q_j$, $\mathbf{J} = \sum_j n_j \, q_j \, \mathbf{u}_j$, and c is the velocity of light.

1.5 PLASMA WAVES

Using the two fluid model (the electrons as one fluid, the ions as the other), we can investigate a wide range of plasma behavior. A characteristic feature of a plasma is its ability to support waves or collective modes of interaction. In the simplest case, these waves correspond to charge density fluctuations at a characteristic frequency determined by the electrons and/or the ions. In a plasma with no large imposed magnetic fields, there are two such plasma waves: a high frequency one called an electron plasma wave and a low frequency one called an ion acoustic wave.

Let us first investigate the high frequency charge density fluctuations associated with the motion of the electrons. Because this is a high frequency oscillation, we can treat the massive ions as an immobile, uniform, neutralizing background with density n_{0i}. Since the wave is electrostatic and the relevant electron motion is along the wave vector (taken to be in the x-direction), a one-dimensional treatment suffices. The equations for an electron fluid with density n_e, mean velocity u_e, and pressure p_e then are

$$\frac{\partial n_e}{\partial t} + \frac{\partial}{\partial x}(n_e\,u_e) = 0 \tag{1.27}$$

$$\frac{\partial}{\partial t}(n_e\,u_e) + \frac{\partial}{\partial x}(n_e\,u_e^2) = -\frac{n_e\,e\,E}{m_e} - \frac{1}{m_e}\frac{\partial p_e}{\partial x} \tag{1.28}$$

$$\frac{p_e}{n_e^3} = \text{constant}. \tag{1.29}$$

We are using the adiabatic equation of state under the assumption that the wave has a phase velocity $\omega/k \gg v_e$, the electron thermal velocity.

It is straightforward to develop an equation for the fluctuations in electron density. First we take a time derivative of Eq. (1.27), a spatial derivative of Eq. (1.28), and eliminate the term $\partial^2 n_e u_e/\partial t\partial x$ to obtain

$$\frac{\partial^2 n_e}{\partial t^2} - \frac{\partial^2}{\partial x^2}(n_e\,u_e^2) - \frac{e}{m}\frac{\partial}{\partial x}(n_e E) - \frac{1}{m_e}\frac{\partial^2 p_e}{\partial x^2} = 0. \tag{1.30}$$

We then use Poisson's equation to relate the electric field to the density:

$$\frac{\partial E}{\partial x} = -4\pi e\left(n_e - Z\,n_{0i}\right), \tag{1.31}$$

where Z is the charge state of the ions.

We next consider small amplitude perturbations in density, velocity and electric field and linearize the equations, i.e., ignore products of the perturbations. If we let $n_e = n_0 + \tilde{n}$, $u_e = \tilde{u}$, $p_e = n_0\theta_e + \tilde{p}$, and $E = \tilde{E}$, Eqs. (1.29-1.31) give

$$\tilde{p} = 3\, m\, v_e^2\, \tilde{n} \tag{1.32}$$

$$\frac{\partial \tilde{E}}{\partial x} = -\, 4\,\pi\, e\, \tilde{n} \tag{1.33}$$

$$\frac{\partial^2 \tilde{n}}{\partial t^2} - \frac{n_0 e}{m}\frac{\partial \tilde{E}}{\partial x} - \frac{\partial^2 \tilde{p}}{\partial x^2} = 0\,. \tag{1.34}$$

Substitution of Eqs. (1.32-1.33) into Eq. (1.34) then gives a wave equation describing the small amplitude fluctuations in electron density:

$$\left(\frac{\partial^2}{\partial t^2} - 3\, v_e^2\, \frac{\partial^2}{\partial x^2} + \omega_{\text{pe}}^2 \right)\tilde{n} = 0\,, \tag{1.35}$$

where $\omega_{\text{pe}} = \sqrt{4\,\pi\, e^2 n_0/m_e}$ is the electron plasma frequency for a plasma with electron density $n_0 = Z\, n_{0i}$. If the density is expressed in units of cm^{-3}, then $\omega_{\text{pe}} = 5.64 \times 10^4\, n_e^{1/2}$.

Looking for a wave-like solution ($\tilde{n} \sim e^{i\,kx - i\,\omega t}$) we readily obtain from Eq. (1.35) the dispersion relation for electron plasma oscillations:

$$\omega^2 = \omega_{\text{pe}}^2 + 3\, k^2 v_e^2\,. \tag{1.36}$$

Note that the frequency of these waves is essentially ω_{pe}, the electron plasma frequency, with a small thermal correction dependent on wavenumber. If kinetic effects are allowed, there will also be a small damping or growth depending on the details of the electron distribution function for velocities near the phase velocity of the wave. This damping will be discussed in Chapter 9.

A plasma will also support charge density oscillations at a much lower frequency determined by the ion inertia. To investigate these oscillations, we need to consider the motion of both the electron and the ion fluids. Since the frequency of these oscillations is much less than the characteristic frequency with which electrons respond (ω_{pe}), we can neglect the inertia of the electrons; i.e., neglect the electron mass. If we again consider motion only along the direction of propagation (taken to be the x direction), the force equation for the electron fluid reduces to

$$n_e\, e\, E = -\, \frac{\partial p_e}{\partial x}\,. \tag{1.37}$$

Since $\omega/k \ll v_e$, the electrons are described by the isothermal equation of state: $p_e = n_e \theta_e$. Substituting p_e into Eq. (1.37) and letting $n_e = n_0 + \tilde{n}_e$ and $E = \tilde{E}$, we obtain the linearized equation:

$$n_0 \, e \, \tilde{E} \; = \; - \, \theta_e \, \frac{\partial \tilde{n}_e}{\partial x} \; , \qquad (1.38)$$

where n_0 is the uniform unperturbed density of the electrons.

The equations for the ion fluid with density n_i, mean velocity u_i, and pressure p_i are

$$\frac{\partial n_i}{\partial t} + \frac{\partial}{\partial x}(n_i u_i) \; = \; 0 \qquad (1.39)$$

$$\frac{\partial}{\partial t}(n_i u_i) + \frac{\partial}{\partial x}(n_i u_i^2) \; = \; \frac{Z\,e}{M} \, n_i \, E - \frac{1}{M}\frac{\partial p_i}{\partial x} \qquad (1.40)$$

$$\frac{p_i}{n_i^3} \; = \; \text{constant} \; ,$$

where Z is the charge state and M the mass of the ions. We use the adiabatic equation of state for the ions under the assumption that $\omega/k \gg v_i$, the ion thermal velocity. To derive an equation for the evolution of the ion density, we proceed as before. Take a time derivative of Eq. (1.39), a spatial derivative of Eq. (1.40), and eliminate $\partial^2 n_i u_i/\partial t \partial x$ to obtain

$$\frac{\partial^2 n_i}{\partial t^2} - \frac{\partial^2}{\partial x^2}(n_i u_i^2) + \frac{Z\,e}{M}\frac{\partial}{\partial x}\left(n_i \, E \right) - \frac{1}{M}\frac{\partial^2 p_i}{\partial x^2} = 0 \; . \qquad (1.41)$$

We now take $n_i = (n_0/Z) + \tilde{n}_i$, $u_i = \tilde{u}_i$, $p_i = p_{i0} + \tilde{p}_i$, and $E = \tilde{E}$, where the superscript denotes small perturbation; further, $\tilde{p}_i = 3\theta_i \tilde{n}_i$, where θ_i is the ion temperature. Substituting these expressions into Eq. (1.41) and neglecting products of the perturbed quantities, we obtain

$$\frac{\partial^2 \tilde{n}_i}{\partial t^2} + \frac{Z\,e\,n_0}{M}\frac{\partial \tilde{E}}{\partial x} - \frac{3\theta_i}{M}\frac{\partial^2 \tilde{n}_i}{\partial x^2} = 0 \; . \qquad (1.42)$$

A wave equation for the fluctuations in ion density is now readily obtained by substituting from Eq. (1.38) into Eq. (1.42) and noting that $\tilde{n}_e \simeq Z\,\tilde{n}_i$, since the electrons closely follow the slow motion of the massive ions:

$$\frac{\partial^2 \tilde{n}_i}{\partial t^2} - \frac{Z\,\theta_e + 3\,\theta_i}{M}\frac{\partial^2 \tilde{n}_i}{\partial x^2} = 0 \; . \qquad (1.43)$$

If we search for wave-like solutions ($\tilde{n}_i \sim e^{i\,kx - i\,\omega t}$), Eq. (1.43) readily gives the dispersion relation for ion acoustic waves:

$$\omega = \pm\, k\, v_s\,, \tag{1.44}$$

where $v_s = \sqrt{(Z\theta_e + 3\theta_i)/M}$ is called the ion-sound velocity. These low frequency waves are the analogue of sound waves in an ordinary gas. The ions provide the inertia, and fluctuations in the pressure provide the restoring force. The electron pressure fluctuations are transmitted to the ions by the electric field. If kinetic effects are allowed, there is a damping due to both the electrons and the ions as will be discussed later in Chapter 9. This damping is small provided that $\omega/k \gg v_i$, which requires that $Z\theta_e \gg \theta_i$. The assumption of quasi-neutrality requires that $k\lambda_{\text{De}} \ll 1$.

1.6 DEBYE SHIELDING

Finally, it is instructive to show how a plasma modifies or shields the electric field of a discrete charge. We place a charge q at rest in a plasma with an initially uniform electron density n_0 and treat the ions as a fixed, neutralizing background. The electrical potential ($\mathbf{E} = -\nabla\phi$) is determined by Poisson's equation:

$$\nabla^2 \phi = -\,4\pi q\,\delta(\mathbf{x}) + 4\pi e\,(n_e - n_0)\,, \tag{1.45}$$

where the charge is located at $r = 0$ for convenience. In the static limit, the force equation for the electron fluid reduces to

$$n_e\, e\, E = -\,\theta_e\,\nabla n_e\,,$$

where an isothermal equation of state has been used i.e., $p_e = n_e\,\theta_e$, where θ_e is the electron temperature. Since $\mathbf{E} = -\nabla\phi$, the electron density then is

$$n_e = n_0\,\exp\!\left(\frac{e\,\phi}{\theta_e}\right). \tag{1.46}$$

Noting that $e\phi/\theta_e \ll 1$, we expand the exponential in Eq. (1.46) and substitute n_e into Eq. (1.45) to obtain

$$\nabla^2 \phi - \frac{\phi}{\lambda_{\text{De}}^2} = -\,4\pi q\,\delta(\mathbf{x})\,, \tag{1.47}$$

where $\lambda_{De} = \sqrt{\theta_e/4\pi n_0 e^2}$ defines the electron Debye length. If θ_e is expressed in units of ev and n_0 in units of cm^{-3}, $\lambda_{De} = 743 (\theta_e/n_0)^{1/2}$. Equation (1.47) is easily solved by Fourier-transforming and then inverting, which gives

$$\phi = \frac{q}{r} \exp\left(\frac{-r}{\lambda_{De}}\right). \tag{1.48}$$

The solution is readily verified by direct substitution. This result demonstrates an important feature of a plasma alluded to earlier in this Chapter. The plasma electrons shield out the field of a discrete charge in a characteristic distance which is λ_{De}. In general, the ions also contribute to the shielding.

References

1. Spitzer, L., *Physics of Fully Ionized Gases*. Interscience Publishers, New York, 1962.

2. Dawson, J. M. and M. B. Gottlieb, *Introduction to Plasma Physics*, Lecture Notes, Princeton University, 1964.

3. Chen, F. F., *Introduction to Plasma Physics*. Plenum Press, New York, 1974.

4. Schmidt, G., *Physics of High Temperature Plasmas*. Academic Press, New York, 1966.

5. Stix, T. H. *The Theory of Plasma Waves*. McGraw-Hill, New York, 1962.

CHAPTER 2

Computer Simulation of Plasmas Using Particle Codes

Having considered a theoretical description in which the plasma is treated as two charged fluids, let us introduce a complementary numerical description of plasma behavior using particle codes [1–12]. Computer simulation of plasma using particle codes is a very direct and powerful approach, particularly for investigating kinetic and/or nonlinear effects. The approach is extremely simple: numerically follow the motion of a large collection of charges in their self-consistent electric and magnetic fields. The basic cycle is illustrated in Fig. 2.1. From the positions and velocities at any given time, compute the charge and current densities on a spatial grid sufficiently fine to resolve the collective behavior. Using these charge and current densities, next compute the self-consistent electric and magnetic fields via Maxwell's equations. Then use these fields in the equations of motion to advance the positions and velocities of the charges. Finally, continue around this basic cycle with a time step sufficiently small to resolve the highest frequency in the problem (which is often the electron plasma frequency).

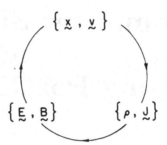

Figure 2.1   The basic cycle of a particle simulation code.

Again, what makes this approach viable is that we are investigating the collective behavior which occurs on a space scale greater than or comparable to the electron Debye length rather than the fine scale fluctuating microfield associated with discrete particle collisions. As we have discussed in Chapter 1, these microfields can be systematically ignored in a collisionless plasma (i.e., when the number of particles in a Debye sphere is much greater than one). This is fortunate since it would be extremely difficult in practice to use a spatial grid fine enough to resolve these microfields, which occur on space scales even less than the interparticle spacing.

Alternatively, we can view our approach as using "finite-size" charges (of size a). It is physically obvious that the behavior of a collection of such charges is the same (with minor modifications) as the behavior of point charges for scale lengths $l \gg a$, but fluctuations with scale lengths $l \ll a$ are suppressed. By choosing $a \sim \lambda_{De}$ (the electron Debye length), we thus achieve with a "trick" what nature does with the use of an enormous number of particles i.e., smooth out the fine scale length microfields.

From a computational viewpoint, a particle code is remarkably simple. The reader can easily write his/her own code. We will first discuss the basic ingredients. Then a very simple code will be presented. This code, and modifications easily made, will be very useful to test theoretical calculations and to investigate various nonlinear effects not readily amenable to analytic treatment.

2.1 BASIC INGREDIENTS OF A PARTICLE CODE

Discussion of a very simple but useful particle code will suffice to illustrate the basic ingredients of all such codes. We treat the ions as a fixed, neutralizing background, assume no imposed magnetic fields, and consider only electrostatic fields. In this electrostatic limit, the magnetic field generated by plasma currents is negligible, and Maxwell's equations reduce to Poisson's equation, $\nabla \cdot \mathbf{E} = 4\pi\rho$. Here $\mathbf{E}$ is the electric field, ρ is the charge density, and cgs units are used. Variations are allowed in only one direction, and periodic boundary conditions are adopted.

Our first task is to compute the charge density. Our system extends from 0 to L, as shown in Fig. 2.2. We divide this system into NC cells and for convenience take the cell size (δ) to be unity (i.e., $L = NC$). The grid points are identified as the integer values of the position, augmented by 1 so that the counting begins with 1. Note that the $NC + 1^{\text{th}}$ grid point is then identical with the first, due to the assumed periodic boundary conditions. Given the position of the charge, many schemes can be used to assign it to the spatial grid. For example, we could just assign the charge to its nearest grid point location. A better scheme is to share the charge between its two nearest grid points. For a charge located a distance Δx to the right of the ith grid point, we then have

$$\Delta\rho(i) = q(1 - \Delta x)$$
$$\Delta\rho(i+1) = q\,\Delta x\,,$$

where $\Delta\rho$ is the increment to the charge density.

Having assigned the charges and determined the charge density on the spatial grid, we next determine the self-consistent electric field using Poisson's equation:

$$\frac{\partial E}{\partial x} = 4\pi\rho\,.$$

The simplest approach is to finite-difference Poisson's equation,

$$E(i+1) = E(i) + 4\pi\delta\left[\frac{\rho(i+1) + \rho(i)}{2}\right].$$

An alternative approach is to Fourier transform the charge density, use Poisson's equation to find E_k ($ik\,E_k = 4\pi\rho_k$), and invert the transform to determine the electric field on the spatial grid.

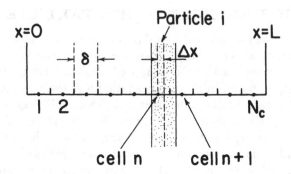

Figure 2.2   The charge sharing scheme for finite-size particles.

The last step in the basic cycle is to use the electric field to move the particles. The electrical force is assigned from the grid to the individual particles using the same scheme chosen to assign the charges to the grid. For example, considering a particle a distance Δx to the right of the ith grid point and using linear interpolation, we obtain for the force F on the particle

$$F = qE(i)[1 - \Delta x] + qE(i+1)\Delta x .$$

The velocity (v) and position (x) of each particle are then advanced Δt in time using a "leap frog" algorithm, i.e.,

$$v^{n+1/2} = v^{n-1/2} + F^n \Delta t$$
$$x^{n+1} = x^n + v^{n+1/2} \Delta t .$$

The superscripts denote time step. By defining x and v one-half time step apart, we achieve second-order accuracy in the time step. In the initial conditions, x and v are defined at the same time $(t = 0)$, but it is straightforward to then displace the velocities backward in time using the force at $t = 0$.

The plasma evolution is computed by simply continuing around the basic cycle using a time step small enough to resolve the characteristic oscillations of the plasma. As discussed in Chapter 1, in this electrostatic limit, the highest frequency oscillation has a frequency near ω_{pe}, the electron plasma frequency. Hence a time step of about $0.2\,\omega_{\text{pe}}^{-1}$ is commonly used. Of course, it is also necessary to resolve the scale lengths characteristic of the collective behavior. Hence a grid size of about λ_{De} is commonly used. It is interesting to note that a numerical instability would be introduced if λ_{De} were chosen to be a small fraction of the grid size [13]. This instability is due to aliasing, which arises from the fact that on the grid a disturbance with wave number k cannot be distinguished from spurious ones with wavenumbers $k + 2\pi n/\delta$, where δ is the grid spacing and n is an integer.

2.2   A 1-D ELECTROSTATIC PARTICLE CODE

As is now apparent, a particle code is quite straightforward from a computational viewpoint. Of course, particle codes become more complex when the full set of Maxwell's equations is allowed and magnetic fields act on the particles. But the basic concepts are the same: use of a spatial grid, the spacing of which is chosen to resolve the collective behavior, and the mapping between the discrete grid and the particles. To further illustrate the ideas, a specific 1-D electrostatic particle code will be presented. For tutorial purposes, the code has intentionally been kept in a form which is easily deciphered. Before presenting this code, it is helpful to briefly discuss two topics: the dimensionless units chosen and the finite-difference solution of Poisson's equation.

The basic equations under consideration are:

$$\frac{dv}{dt} = -\frac{eE}{m}$$

$$\frac{dx}{dt} = v$$

$$\frac{\partial E}{\partial x} = -4\pi e\left(n - n_{av}\right),$$

where n_{av} is the density of the fixed ion background and n is the density of the simulation electrons with charge e and mass m. We choose dimensionless variables in the following way:

$$t' = \omega_{pe}\,t$$

$$x' = \frac{x}{\delta}$$

$$v' = \frac{v}{\omega_{pe}\,\delta}$$

$$E' = \frac{E}{4\pi\,e\,n_{av}\delta}\,,$$

where $\delta = L$ (system size)$/NC$ (the number of cells), $n_{\mathrm{av}} = NP$ (the number of simulation particles)$/L$, and $\omega_{\mathrm{pe}} = (4\pi\, n_{\mathrm{av}}\, e^2/m)^{1/2}$. In these variables, the equations become

$$\frac{dv'}{dt'} = -E'$$

$$\frac{dx'}{dt'} = v'$$

$$\frac{dE'}{dx'} = -\frac{1}{N_{\mathrm{av}}}\,[\, N - N_{\mathrm{av}}\,]\ .$$

Here $N = n\delta$ is the number of particles initially assigned to each cell and $Nav = n_{\mathrm{av}}\delta = NP/NC$. In terms of these variables, the total energy (TE) of the system is

$$TE = \sum_{i=1}^{NP} \frac{m}{2}\, v_i^2 + \sum_{i=1}^{NC} \frac{E_i^2}{8\pi}\, \delta$$

$$TE = \frac{m\,\omega_{\mathrm{pe}}^2\,\delta^2}{2}\left[\sum_{i=1}^{NP} v_i'^2 + N_{\mathrm{av}} \sum_{i=1}^{NC} E_i'^2\right]\ .$$

The finite-difference solution to Poisson's equation is also straightforward. Defining a normalized charge density RH, we have

$$\frac{\partial E'}{\partial x'} = RH\ .$$

After the charges are assigned to the grid, we know $RH(i)$, $(i = 1, NC)$. Dropping the prime notation and taking $\delta = 1$ then gives

$$E\,(i+1) = E\,(i) + \frac{1}{2}\,[RH(i+1) + RH(i)]\ .$$

Considering only periodic boundary conditions imposes the constraint that $E(NC+1) = E(1)$.

Our procedure is to first solve for $E(1)$, using the condition that the spatially-averaged electric field is zero i.e.,

$$\sum_{j=1}^{NC} E(j) = 0\ .$$

For $j > 1$,

$$E(j) = E(1) + \sum_{i=1}^{j-1} \frac{1}{2} [RH(i) + RH(i+1)] .$$

Substituting, we obtain

$$E(1) + \sum_{j=2}^{NC} \left\{ E(1) + \sum_{i=1}^{j-1} \frac{1}{2} [RH(i) + RH(i+1)] \right\} = 0 .$$

Hence

$$NC \times E(1) = - \sum_{j=2}^{NC} \sum_{i=1}^{j-1} \frac{1}{2} [RH(i) + RH(i+1)] .$$

Since j is simply a counting variable, we let $j' = j - 1$ to obtain

$$\sum_{j=2}^{NC} \sum_{i=1}^{j-1} = \sum_{j'=1}^{NC-1} \sum_{i=1}^{j'} .$$

Note that we can add the $j' = NC$ term since by charge neutrality the spatially-averaged charge density vanishes ($\sum_{i=1}^{NC} RH(i) = 0$). Hence

$$NC \times E(1) = - \sum_{j'=1}^{NC} \sum_{i=1}^{j'} \frac{1}{2} [RH(i) + RH(i+1)] .$$

With $E(1)$ determined, each of the $E(i)$ is readily found.

A Fortran implementation of the code is presented in Fig. 2.3. The coding is self-explanatory; many comment cards are included. Arrays x and v have dimensions corresponding to the number of particles NP and arrays RH and E have dimensions corresponding to the number of cells NC. For clarity of presentation, various optimization techniques, such as time normalization in units of Δt, have not been exploited. The cells to which a given particle is assigned are determined by truncation of its position using fixed-point arithmetic conversion. Prior to entering the basic cycle, the particles are initialized to represent the initial state of the plasma for the problem under consideration. The most common diagnostics are the temporal evolution of the electrostatic and kinetic energies,

```
PROGRAM RAIN(INPUT,HSP)
CALL KEEP88(1)
C ONE DIMENSIONAL PARTICLE SIMULATION
C PERIODIC BOUNDARY CONDITIONS
C FIXED IONS
C NC IS THE NUMBER OF CELLS
C NP IS THE NUMBER OF ELECTRONS
C THESE ARRAYS ARE OF SIZE NP
      DIMENSION X(10000),V(10000)
C THESE ARRAYS ARE OF SIZE NC (PLUS 1)
      DIMENSION RH(300),E(300)
C THESE ARRAYS ARE OF SIZE MEND
      DIMENSION EE(1000),EK(1000),TE(1000)
C ARRAY OF SIZE MEND OR NC
      DIMENSION PL(1000)
C GIVE INPUT DATA
      WRITE(59,21)
   21 FORMAT(" GIVE NC,NP,MP IN 3I5")
      READ(59,22) NC,NP,MP
   22 FORMAT (3I5)
      WRITE(59,23) NC,NP,MP
   23 FORMAT(" NC = ",I5/" NP = ",
     1I5/" MP = ",I5)
      WRITE (59,24)
   24 FORMAT(" GIVE VTE,DT,TEND IN F9.5")
      READ(59,25) VTE,DT,TEND
   25 FORMAT (3F9.5)
      WRITE(59,26) VTE,DT,TEND
   26 FORMAT(" VTE = ",F9.5/" DT = ",F9.5/
     1" TEND = ",F9.5)
      CALL SETCH(1.,40.,1,0,1,0,0)
      WRITE(100,23) NC,NP,MP
      WRITE(100,26) VTE,DT,TEND
      CALL FRAME
      ANC=NC
      NC1=NC+1
      ANP=NP
      ANORM=ANC/ANP
      T=0.
      K1=0
      L1=0
      M1=0
      NDT=1./DT
      ND1=NDT-1
      MEND=TEND+0.1
C SET PLOTTING ARRAY
      MPLOT=MEND
      IF (NC.GT.MPLOT) MPLOT=NC
      DO 3001 I=1,MPLOT
      AI=I
 3001 PL(I)=AI
      AK1=6.2832/ANC
C INITIALIZE THE PARTICLES
C MAKE SURE PARTICLES IN SYSTEM
      DO 1 I=1,NP
      AI=I-.5
      X(I)=AI*ANORM
      IF (X(I).LT.0.) X(I)=X(I)+ANC
      IF (X(I).GE.ANC) X(I)=X(I)-ANC
    1 CONTINUE
      VB=0.
      DO 2 I=1,NP
      VB=-VB
      CALL RANDUM(VV)
    2 V(I)=VTE*VV+VB
C INITIALLY ASSIGN THE CHARGE TO THE GRID
      DO 3 I=1,NC1
    3 RH(I)=0.
      DO 4 I=1,NP
      N=X(I)
      AN=N
      DX=X(I)-AN
      N=N+1
      M=N+1
      RH(N)=RH(N)-1.*(1.-DX)
      RH(M)=RH(M)-DX
    4 CONTINUE
    5 CONTINUE
      RH(1)=RH(1)+RH(NC1)
C ADD IN CHARGE OF UNIFORMLY SPACED IONS
      DO 6 I=1,NC
      RH(I)=RH(I)+ANP/ANC
    6 RH(I)=RH(I)*ANORM
C SOLVE POISSONS EQUATION
C FIND E(1) --AVERAGE FIELD IS ZERO
      ERUN2=0.
      DO 32 J=1,NC
      ERUN1=0.
      DO 31 I=1,J
      K=I+1
      IF (K.GT.NC) K=1
   31 ERUN1=ERUN1+0.5*(RH(I)+RH(K))
   32 ERUN2=ERUN2+ERUN1
      E(1)=-ERUN2/ANC
      NCM=NC-1
      DO 7 I=1,NCM
    7 E(I+1)=E(I)+0.5*(RH(I+1)+RH(I))
      E(NC+1)=E(1)
C ZERO THE CHARGE DENSITY ARRAY
      DO 8 I=1,NC1
    8 RH(I)=0.
C COMPUTE KINETIC AND FIELD ENERGIES
      IF (K1.LT.ND1) GO TO 41
      EEA=0.
      DO 42 I=1,NC
   42 EEA=EEA+E(I)*E(I)
      EEA=EEA/ANORM
      EK1=0.
      DO 43 I=1,NP
   43 EK1=EK1+V(I)*V(I)
   41 CONTINUE
C AT T=0., USE ONLY HALF OF E
C TO STAGGER THE VELOCITIES
      IF(T.NE.0.) GO TO 51
      DO 50 I=1,NC1
   50 E(I)=0.5*E(I)
   51 CONTINUE
C PARTICLE LOOP
C MOVE THE ELECTRONS
      DO 9 I=1,NP
      N=X(I)
      AN=N
      DX=X(I)-AN
      N=N+1
      M=N+1
      FORCE=-E(N)*(1.-DX)-E(M)*DX
      V(I)=V(I)+FORCE*DT
      X(I)=X(I)+V(I)*DT
C MAKE SURE PARTICLE IS WITHIN THE SYSTEM
      IF (X(I).LT.0.) X(I)=X(I)+ANC
      IF (X(I).GE.ANC) X(I)=X(I)-ANC
C ASSIGN CHARGE TO THE GRID--AREA WEIGHTING
      N=X(I)
      AN=N
      DX=X(I)-AN
      N=N+1
      M=N+1
      RH(N)=RH(N)-(1.-DX)
      RH(M)=RH(M)-DX
    9 CONTINUE
      T=T+DT
      K1=K1+1
      IF (K1.LT.NDT) GO TO 5
      K1=0
      L1=L1+1
C COMPUTE THE KINETIC ENERGY
      EK2=0.
      DO 44 I=1,NP
   44 EK2=EK2+V(I)*V(I)
C SAVE THE ENERGIES FOR LATER PLOTTING
      EK(L1)=0.5*(EK1+EK2)
      EE(L1)=EEA
      TE(L1)=EK(L1)+EE(L1)
      M1=M1+1
C PLOT SNAPSHOTS OF THE SYSTEM EVERY WPE*T=MP
C SUBSTITUTE YOUR LOCAL PLOTTING ROUTINES
      IF (M1.LT.MP) GO TO 10
      M1=0
      CALL CARTMM(NC,YMIN,YMAX,E,1)
      CALL MAPG(0.,ANC,YMIN,YMAX)
      CALL SETCH(0.,2.,0,0,1,0,0)
      WRITE(100,61) T
   61 FORMAT(" E FIELD VS POSITION "/
     1" TIME = ",F10.5)
      CALL TRACE(PL,E,NC,1,1)
      CALL FRAME
      VMIN=-12.*VTE
      VMAX=12.*VTE
      CALL MAPS(0.,ANC,VMIN,VMAX)
      CALL SETCH(0.,2.,0,0,1,0,0)
      WRITE(100,62) T
   62 FORMAT(" PHASE SPACE "/
     1" TIME = ",F10.5)
      CALL POINTS(X,V,NP,1,1)
      CALL FRAME
C PLOT TIME HISTORIES AND OTHER FINAL PLOTS
C SUBSTITUTE YOUR LOCAL PLOTTING ROUTINES
   10 IF (T.LT.TEND) GO TO 5
      CALL CARTMM(MEND,YMIN,YMAX,EE,1)
      CALL MAPG(0.,TEND,YMIN,YMAX)
      CALL SETCH(0.,2.,0,0,1,0,0)
      WRITE(100,63) T
   63 FORMAT(" E FIELD ENERGY VS TIME "/
     1" TIME = ",F10.5)
      CALL TRACE(PL,EE,MEND,1,1)
      CALL FRAME
      CALL CARTMM(MEND,YMIN,YMAX,EK,1)
      CALL MAPG(0.,TEND,YMIN,YMAX)
      CALL SETCH(0.,2.,0,0,1,0,0)
      WRITE(100,64) T
   64 FORMAT(" KINETIC ENERGY VS TIME "/
     1" TIME = ",F10.5)
      CALL TRACE(PL,EK,MEND,1,1)
      CALL FRAME
      CALL CARTMM(MEND,YMIN,YMAX,TE,1)
      CALL MAPG(0.,TEND,YMIN,YMAX)
      CALL SETCH(0.,2.,0,0,1,0,0)
      WRITE(100,65) T
   65 FORMAT(" TOTAL ENERGY VS TIME "/
     1" TIME = ",F10.5)
      CALL TRACE(PL,TE,MEND,1,1)
      CALL FRAME
      CALL EXIT
      END
C COMPUTE RANDOM NUMBERS
C WITH A GAUSSIAN DISTRIBUTION
      SUBROUTINE RANDUM(X1)
      X1=0.
      DO 15 I=1,12
   15 X1=X1+RNFL(DD)-.5
      RETURN
      END
*DATA
```

Figure 2.3   Listing of a one-dimensional particle code.

snapshots of the electric field as a function of position, and plots of the $x - v$ phase space of the particles at various times. Finally, we note that energy conservation is an important test of the code performance.

Even with this simplest particle code, many instructive problems can be investigated. For example, we can excite electron plasma waves and examine their behavior in both the linear and nonlinear regimes. The code can be easily extended to include ion motion, and the reader is encouraged to do this as a learning experience. After becoming familiar with the basic techniques, one can easily graduate to the use of more flexible and optimized particle codes, such as **ES1** which is available through NMFECC, the national computer center for magnetic fusion energy. Although less accessible, large 2 and 2-$\frac{1}{2}$ dimensional (two-dimensions in space and three in velocity) electromagnetic, relativistic particle codes now exist. Such codes solve the full set of Maxwell's equations and use relativistic particle dynamics. So-called implicit particle codes have also been constructed [14]. These codes allow time steps greater than ω_{pe}^{-1} and are used to more economically study low frequency kinetic phenomena.

References

1. Introductory articles:
   Denavit, J. and W. L. Kruer, How to get started in particle simulation, *Comments on Plasma Physics and Controlled Fusion* **6**, 35 (1980);
   W. L. Kruer, Plasma simulation using particle codes, *Nuclear Technology* **27**, 216 (1975).

2. Birdsall, C. K. and A. B. Langdon, *Plasma Physics via Computer Simulation.* McGraw-Hill, New York 1985.

3. Hockney, R. W. and J. W. Eastwood, *Computer Simulation Using Particles.* McGraw-Hill, New York 1981.

4. Alder, B., S. Fernback, and M. Rotenburg, eds., *Methods in Computational Physics*, Vols. 9 and 16. Academic Press, New York, 1970 and 1976.

5. Buneman, O., Dissipation currents in ionized media, *Phys. Rev.* **115**, 503 (1959).

6. Dawson, J. M., One-dimensional plasma model, *Phys. Fluids* **5**, 445 (1962).

7. Hockney, R. W., Computer experiment of anomalous diffusion, *Phys. Fluids* **9**, 1826 (1966).

8. Morse, R. L. and C. W. Nielsen, One-, two-, and three-dimensional simulation of two-beam plasmas, *Phys. Rev. Letters* **23**, 1087 (1969).

9. Birdsall, C. K. and D. Fuss, Clouds-in-clouds, clouds-in-cells physics for many-body plasma simulation, *J. Comp. Phys.* **3**, 494 (1969).

10. Boris, J. P. and K. V. Roberts, The optimization of particle calculations in 2 and 3 dimensions, *J. Comp. Phys.* **4**, 552 (1969).

11. Kruer, W. L., J. M. Dawson, and B. Rosen, Dipole expansion method for plasma simulation, *J. Comp. Phys.* **13**, 114 (1973).

12. Denavit, J. and W. L. Kruer, Comparison of numerical solutions of the Vlasov equation with particle simulations of collisionless plasmas, *Phys. Fluids* **14**, 1782 (1971).

13. Langdon, A. B., Effects of the spatial grid in simulation plasmas, *J. Comp. Phys.* **6**, 247 (1970).

14. For several articles on implicit particle codes, see *Multiple Time Scales*, (Brackbill, J. U., and B. I. Cohen, eds.). Academic Press, Orlando, 1985.

CHAPTER 3

Electromagnetic Wave Propagation in Plasmas

3.1 WAVE EQUATION FOR LIGHT WAVES IN A PLASMA

Having examined the characteristic charge density oscillations which are supported by a plasma, let us now consider how a plasma modifies the propagation of electromagnetic waves. Motivated by laser fusion applications, we will continue to assume that there are no large imposed (or self-generated) magnetic fields. We begin by examining the linearized plasma response to a high frequency field of the form [1]

$$\mathbf{E} = \mathbf{E}(\mathbf{x}) \exp(-i\,\omega t) \,. \tag{3.1}$$

Since the frequency ω is assumed to be $\gtrsim \omega_{\mathrm{pe}}$, the ions are treated as a stationary, neutralizing background with density $n_{0i}(\mathbf{x})$. If we neglect the terms involving $\mathbf{u}_e \cdot \nabla \mathbf{u}_e$ and $\mathbf{u}_e \times \mathbf{B}$ as products of small quantities (i.e., $\sim |\mathbf{E}|^2$), the linearized force equation for the velocity of the electron fluid reduces to

$$\frac{\partial \mathbf{u}_e}{\partial t} = -\frac{e}{m} \, \mathbf{E}(\mathbf{x}) \, \exp(-i\,\omega t) \, .$$

Since the current density is $J = -n_0(\mathbf{x})\, e\, \mathbf{u}_e$,

$$\frac{\partial\, \mathbf{J}}{\partial t} = -n_0(\mathbf{x})\, e\, \frac{\partial \mathbf{u}_e}{\partial t} = \frac{\omega_{\mathrm{pe}}^2(\mathbf{x})}{4\,\pi}\, \mathbf{E} \, ,$$

where $\omega_{\mathrm{pe}}^2 = 4\,\pi\, e^2 n_0/m$ and $n_0 = Z\, n_{0i}$ is the electron density. Hence

$$\mathbf{J} = \frac{i\,\omega_{\mathrm{pe}}^2}{4\,\pi\,\omega}\, \mathbf{E} = \sigma\, \mathbf{E} \, ,$$

where the high frequency conductivity of the plasma is $\sigma = i\omega_{\mathrm{pe}}^2/4\,\pi\,\omega$.

To develop wave equations for the oscillating electric and magnetic fields, we first consider Faraday's law and Ampere's law, which become

$$\nabla \times \mathbf{E} = \frac{i\,\omega}{c}\, \mathbf{B} \, , \tag{3.3}$$

$$\nabla \times \mathbf{B} = \frac{4\,\pi}{c}\, \sigma\, \mathbf{E} - \frac{i\,\omega}{c}\, \mathbf{E} \, . \tag{3.4}$$

Substituting for σ into Eq. (3.4) gives

$$\nabla \times \mathbf{B} = -\frac{i\,\omega}{c}\, \epsilon\, \mathbf{E} \, , \tag{3.5}$$

where $\epsilon = 1 - \omega_{\mathrm{pe}}^2/\omega^2$ defines the dielectric function of the plasma. Taking the curl of Eq. (3.3), substituting from Eq. (3.5), and using a standard vector identity gives

$$\nabla^2 \mathbf{E} - \nabla(\nabla \cdot \mathbf{E}) + \frac{\omega^2}{c^2}\, \epsilon\, \mathbf{E} = 0 \, . \tag{3.6}$$

The wave equation for $\mathbf{B}$ is developed in a similar fashion. The curl of Eq. (3.5) gives

$$\nabla \times (\nabla \times \mathbf{B}) = -\frac{i\,\omega}{c}\, \nabla \times (\epsilon\, \mathbf{E}) \, .$$

Since $\nabla \times \epsilon \mathbf{E} = \epsilon \nabla \times \mathbf{E} + \nabla \epsilon \times \mathbf{E}$, we obtain

$$\nabla^2 \mathbf{B} + \frac{\omega^2}{c^2} \epsilon \mathbf{B} + \frac{1}{\epsilon} \nabla \epsilon \times (\nabla \times \mathbf{B}) = 0 . \qquad (3.7)$$

As our first application of these results, let us derive the dispersion relation for electromagnetic waves in a plasma with a uniform density. Since $\nabla \epsilon = 0$ and $\nabla \cdot \mathbf{E} = 0$, the wave equations for $\mathbf{E}$ and $\mathbf{B}$ become identical. Assuming a spatial dependence of $\exp(i\mathbf{k} \cdot \mathbf{x})$ then gives the dispersion relation for electromagnetic waves in a plasma:

$$\frac{\omega^2}{c^2} \epsilon = k^2$$

or

$$\omega^2 = \omega_{pe}^2 + k^2 c^2 . \qquad (3.8)$$

Note that ω_{pe} is the minimum frequency for propagation of a light wave in a plasma i.e., k becomes imaginary for $\omega < \omega_{pe}$. Since the characteristic response time for electrons is ω_{pe}^{-1}, the electrons shield out the field of a light wave when $\omega < \omega_{pe}$. Hence the condition $\omega_{pe} = \omega$ defines the maximum plasma density to which a light wave can penetrate. This so-called critical density is $n_{cr} = 1.1 \times 10^{21}/\lambda_\mu^2$ cm^{-3}, where λ_μ is the free-space wavelength of the light in units of microns ($1\mu m = 10^{-4}$ cm).

To investigate some of the basic features of the propagation of light waves in an inhomogeneous plasma, let's consider plane waves normally incident onto a plasma slab. Assuming variations only in the z direction, we then have

$$n_0 = n_0(z)$$
$$\epsilon = \epsilon(\omega, z)$$
$$\mathbf{E}(\mathbf{x}) = \mathbf{E}(z) \exp(-i\omega t) . \qquad (3.9)$$

In Cartesian coordinates, the wave equation for $\mathbf{E}$ (Eq. 3.6) becomes

$$\frac{d^2}{dz^2} E_{x,y} + \frac{\omega^2}{c^2} \epsilon E_{x,y} = 0 ,$$
$$\epsilon E_z = 0 . \qquad (3.10)$$

Likewise, the wave equation for **B** (Eq. 3.7) reduces to

$$\frac{d^2}{dz^2}\,B_{x,y} + \frac{\omega^2}{c^2}\,\epsilon\,B_{x,y} - \frac{1}{\epsilon}\frac{d\epsilon}{dz}\frac{dB_{x,y}}{dz} = 0\,,$$

$$\frac{d\,B_z}{dz} = 0\,.$$

We will first develop a WKB solution for the fields. Although limited to weak density gradients, this solution provides an excellent illustration of how gradients in the density affect the wave propagation. Then we will complement this analysis with an exact solution for the fields, assuming a linear variation in the plasma density.

3.2 WKB SOLUTION FOR WAVE PROPAGATION IN AN INHOMOGENEOUS PLASMA

A very useful approximate solution for the wave propagation can be obtained in the limit that the fields vary slowly in space. It's most convenient to solve for the electric field. If we take the electric vector to be in the x direction and let $E_x = E$, Eq. (3.10) becomes

$$\frac{d^2 E}{dz^2} + \frac{\omega^2}{c^2}\,\epsilon\,(\omega,z)\,E = 0\,. \tag{3.11}$$

We assume a slow variation in the dielectric function of the plasma (i.e., a weak gradient in density) and look for a solution of the form

$$E(z) = E_0(z)\,\exp\left[\frac{i\,\omega}{c}\int^z \Psi(z')\,dz'\right]\,, \tag{3.12}$$

where the amplitude $E_0(z)$ and the phase $\Psi(z)$ are slowly varying functions. Differentiating $E(z)$ and substituting into Eq. (3.11) yields

$$E_0'' + \frac{2i\omega}{c}\Psi\,E_0' - \frac{\omega^2}{c^2}\Psi^2\,E_0 + \frac{i\omega}{c}\Psi'\,E_0 + \frac{\omega^2}{c^2}\,\epsilon\,E_0 = 0\,,$$

where the prime denotes a derivative with respect to z. To lowest order, we neglect all derivatives, obtaining

$$\Psi = \sqrt{\epsilon(\omega,z)}\,. \tag{3.13}$$

To next order in the gradients,

$$\frac{2\,i\,\omega}{c}\,\Psi\,E_0' + \frac{i\,\omega\,\Psi'\,E_0}{c} = 0 \,, \qquad (3.14)$$

with the solution

$$E_0(z) = \frac{\text{constant}}{\sqrt{\Psi}} \,. \qquad (3.15)$$

Substituting Eqs. (3.13) and (3.15) into (3.12) then gives

$$E(z) = \frac{E_{\text{FS}}}{\epsilon^{1/4}}\,\exp\left[\frac{i\omega}{c}\int^z \sqrt{\epsilon(\omega, z')}\,\mathrm{d}z'\right] \,, \qquad (3.16)$$

where E_{FS} is the value of the electric field in free space.

It is apparent from Eq. (3.16) that the amplitude of the electric field increases as the light wave propagates toward higher density. This behavior is readily explained physically by noting that the energy flux is conserved, i.e.,

$$\frac{v_g\,|E(z)|^2}{8\,\pi} = \frac{c\,E_{\text{FS}}^2}{8\,\pi} \,, \qquad (3.17)$$

where v_g is the group velocity of the light wave in the plasma. Using the dispersion relation (Eq. 3.8) to relate the local value of the group velocity to the local value of the dielectric function gives $v_g/c = \sqrt{\epsilon(\omega, z)}$. Hence, Eq. (3.17) becomes

$$|E_0(z)| = \frac{E_{\text{FS}}}{\epsilon^{1/4}} \,,$$

in agreement with the WKB result. Since the energy flux can also be expressed as $c\,\mathbf{E}\times\mathbf{B}/4\,\pi$, we can easily see that the amplitude B_0 of the magnetic field of a light wave is decreased in a plasma, i.e.,

$$|B_0(z)| = B_{\text{FS}}\,\epsilon^{1/4} \,,$$

where B_{FS} is the value of the magnetic field in vacuum.

The validity condition for the WKB solution can be readily estimated. For example, in deriving Eq. (3.14), we required that

$$E_0'' \ll \frac{\omega}{c}\,\Psi'\,E_0 \,, \frac{\omega}{c}\,\Psi\,E_0' \,.$$

Noting from Eq. (3.16) that $k(z) = \omega\Psi/c$, it suffices to require that

$$\frac{\partial E_0}{\partial z} \ll k\,E_0 \,.$$

Substituting for E_0 from Eq. (3.16), we then obtain the condition

$$\lambda\, \frac{\partial \epsilon}{\partial z} \ll 8\pi\,\epsilon \,, \qquad (3.18)$$

where $\lambda(z) = 2\pi/k(z)$. In other words, the variation in the plasma density must be sufficiently slow that the fractional change in the dielectric function in a local wavelength is very small. Note that the WKB solution breaks down near the critical density where $\epsilon \to 0$ and $\lambda \to \infty$.

3.3   ANALYTIC SOLUTION FOR PLASMA WITH A CONSTANT DENSITY GRADIENT

The WKB solution is especially valuable because it provides us with a very intuitive way to describe the wave propagation. The wave is assigned a wavenumber and group velocity which are defined via the local dispersion relation. The amplitude and the phase of the wave are related to these locally-defined characteristics in a physically obvious way. However, strictly speaking, this approximation is valid only for very gentle density gradients and in particular breaks down near the cutoff density where the wave reflects. Hence it's important to complement the WKB solution with a more complete solution of the wave propagation. Fortunately, an exact solution can readily be obtained for a plasma with a linear variation in density [2].

We again consider a plane electromagnetic wave normally incident onto an inhomogeneous plasma slab and allow variations only in the z direction. The wave equation for the electric field $E(z)$ is

$$\frac{\mathrm{d}^2 E}{\mathrm{d}z^2} + \frac{\omega^2}{c^2}\, \epsilon(\omega, z)\, E = 0 \,.$$

Assuming that the plasma density is a linear function of position ($n = n_{\mathrm{cr}}\, z/L$, where $n_{\mathrm{cr}} = m\omega^2/4\pi e^2$ is the critical density), we obtain

$$\frac{\mathrm{d}^2 E}{\mathrm{d}z^2} + \frac{\omega^2}{c^2}\left(1 - \frac{z}{L}\right) E = 0 \,. \qquad (3.19)$$

A change of variables to $\eta = (\omega^2/c^2 L)^{1/3}(z - L)$ gives

$$\frac{\mathrm{d}^2 E}{\mathrm{d}\eta^2} - \eta\, E = 0 \,. \qquad (3.20)$$

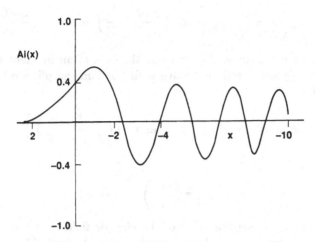

Figure 3.1   A plot of the Airy function $A_i(x)$.

This differential equation defines the well-documented Airy functions, A_i and B_i [3]. The general solution of Eq. (3.20) is

$$E(\eta) = \alpha\, A_i(\eta) + \beta\, B_i(\eta)\,,$$

where α and β are constants which are determined by matching to the boundary conditions.

On physical grounds, we expect E to represent a standing wave for $\eta < 0$ and to decay as $\eta \to \infty$. Since $B_i(\eta) \to \infty$ as $\eta \to \infty$, we choose $\beta = 0$. A plot of $A_i(\eta)$ is shown in Fig. 3.1. Note that the wavelength and the amplitude of the field increases as the reflection point ($\eta = 0$) is approached, as expected from the WKB solution. Beyond the reflection point, the field is attenuated.

The constant α is chosen by matching the electric field with the field of the incident light wave at the interface between the vacuum and the plasma at $z = 0$ i.e., at $\eta = -(\omega L/c)^{2/3}$. If we assume that $\omega L/c \gg 1$ and use the asymptotic representation,

$$A_i(-\eta) = \frac{1}{\sqrt{\pi}\,\eta^{1/4}} \cos\left(\frac{2}{3}\eta^{2/3} - \frac{\pi}{4}\right)\,,$$

we obtain

$$E(z\!=\!0) = \frac{\alpha}{2\sqrt{\pi}\,(\omega L/c)^{1/6}} \left[\exp\, i\!\left(\frac{2}{3}\frac{\omega L}{c} - \frac{\pi}{4}\right) + \exp\, -i\!\left(\frac{2}{3}\frac{\omega L}{c} - \frac{\pi}{4}\right) \right].$$

Note that we can express $E(z = 0)$ as the sum of an incident wave with amplitude E_{FS} and a reflected wave with the same amplitude but shifted in phase i.e.,

$$E(z\!=\!0) = E_{\text{FS}} \left[1 + \exp\, -i\left(\frac{4}{3}\frac{\omega L}{c} - \frac{\pi}{2}\right) \right],$$

provided

$$\alpha = 2\sqrt{\pi} \left(\frac{\omega L}{c}\right)^{1/6} E_{\text{FS}}\, e^{i\varphi}.$$

Here E_{FS} is the free-space value of the electric field of the incident light wave and φ is just a phase factor which does not affect $|E|$. Hence

$$E(\eta) = 2\sqrt{\pi} \left(\frac{\omega L}{c}\right)^{1/6} E_{\text{FS}}\, e^{i\varphi}\, A_i(\eta). \qquad (3.21)$$

As can be seen from Fig. 3.1, the amplitude of the electric field reaches a maximum value at $\eta = 1$, which corresponds to $(z - L) = -(c^2 L/\omega^2)^{1/3}$. This maximum amplitude (E_{max}) is

$$\left|\frac{E_{\text{max}}}{E_{\text{FS}}}\right|^2 \simeq 3.6 \left(\frac{\omega L}{c}\right)^{1/3}. \qquad (3.22)$$

We would expect a factor of four increase in E^2 because a standing wave is set up. The additional swelling is due to the decrease in the group velocity of the light wave as the dielectric function becomes small.

A similar swelling of the peak electric field amplitude can be obtained by heuristic arguments based on WKB theory. Here we use $k = \sqrt{\epsilon}\,(\omega/c)$ and $|E| = E_{\text{FS}}/\epsilon^{1/4}$. As ϵ becomes smaller, the wavelength becomes longer. Intuitively we expect the wave to average over the plasma properties within at least half of its local wavelength. Hence we expect a minimum value of ϵ, which is roughly the value of ϵ averaged over half a local wavelength near the reflection point. For a linear density profile, $\epsilon_{\text{min}} \approx \pi/(2k_{\text{min}}L)$. Since $k_{\text{min}} = \epsilon_{\text{min}}^{1/2}\, \omega/c$, we then obtain $\epsilon_{\text{min}} \approx (\pi c/2\omega L)^{2/3}$. Including the factor of 2 in amplitude which is due to the standing wave, we then estimate $|E_{max}/E_{FS}|^2 \approx 3.5(\omega L/c)^{1/3}$, a value in reasonable agreement with the exact solution. It's also interesting that the phase shift between the incident and reflected waves is given by the WKB solution if we simply subtract $\pi/2$ to account for reflection at the critical density surface. In other words,

$$\Psi = 2\,\frac{\omega}{c} \int_0^L \sqrt{\epsilon}\,dz - \frac{\pi}{2} = \frac{4}{3}\frac{\omega L}{c} - \frac{\pi}{2}\,.$$

The magnetic field of the light wave is readily calculated from the solution for E. Noting that the electric vector is in the x direction and that the wave is propagating in the z direction, we take the y-component of Faraday's law to obtain

$$B = -\frac{i\,c}{\omega}\frac{\partial E}{\partial z}\,.$$

Changing variables from z to η and using Eq. (3.21) gives

$$B(\eta) = -i\,2\,\sqrt{\pi}\left(\frac{c}{\omega L}\right)^{1/6} E_{FS}\,e^{i\varphi}\,A'_i(\eta)\,, \qquad (3.23)$$

where the prime denotes a derivative with respect to η. At the reflection point, $|B(\eta = 0)| \approx 0.92\,(c/\omega L)^{1/6}\,E_{FS}$. Note that B decreases as E swells, as qualitatively shown by the WKB solution.

References

1. Jackson, J. D., *Classical Electrodynamics*. Wiley, New York, 1962.

2. Ginzberg, V. L., *The Properties of Electromagnetic Waves in Plasmas*. Pergamon, New York, 1964.

3. *Handbook of Mathematical Functions* (Abramowitz, M. and I. A. Stegun, eds.). National Bureau of Standards, *Applied Mathematics Series* **55**, 1964.

CHAPTER 4

Propagation of Obliquely Incident Light Waves in Inhomogeneous Plasmas

To complete our introduction to the propagation of light waves in inhomogeneous plasmas, let us now consider a light wave whose propagation vector is at an angle to the density gradient. We again consider a plane electromagnetic wave incident onto a plasma slab with electron density $n_e(z)$. The vacuum-plasma interface is taken to be at $z = 0$, where the angle of incidence θ is defined as the angle between the propagation vector $\mathbf{k}$ and the direction of the density gradient ($\hat{z}$). Without loss of generality, we take the plane of incidence (defined by the vectors ∇n and $\mathbf{k}$) to be the $y - z$ plane, as shown in Fig. 4.1. With this choice, there are no variations in the x-direction (i.e., $k_x = 0$ and $\frac{\partial}{\partial x} = 0$). At the vacuum-plasma interface, we note that $k_y = (\omega/c) \sin \theta$ and $k_z = (\omega/c) \cos \theta$. As we will show, the wave propagation now depends on whether the electric vector $\mathbf{E}$ of the incident light wave lies *in* or *out* of the plane of incidence.

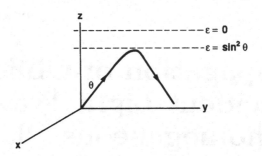

Figure 4.1   A sketch illustrating a light ray obliquely incident onto an inhomogeneous plasma slab.

4.1   OBLIQUELY INCIDENT S-POLARIZED LIGHT WAVES

If the electric vector points out of the plane of incidence, the light wave is termed s-polarized. If we then take $\mathbf{E} = E_x \hat{x}$, the wave equation for $\mathbf{E}$ (Eq. 3.6) becomes

$$\frac{\partial^2 E_x}{\partial y^2} + \frac{\partial^2 E_x}{\partial z^2} + \frac{\omega^2}{c^2}\,\epsilon(z)\,E_x = 0 \,. \tag{4.1}$$

Since the dielectric function (ϵ) of the plasma is a function of z alone, k_y must be conserved. Hence $k_y = (\omega/c)\sin\theta$ and

$$E_x = E(z)\,\exp\left(\frac{i\,\omega\,y\,\sin\theta}{c}\right), \tag{4.2}$$

where θ is the angle of incidence. Substituting Eq. (4.2) into Eq. (4.1) then gives

$$\frac{\mathrm{d}^2 E(z)}{\mathrm{d}z^2} + \frac{\omega^2}{c^2}\left(\epsilon(z) - \sin^2\theta\right)E(z) = 0 \,. \tag{4.3}$$

It is apparent that reflection of the light wave now occurs when

$$\epsilon(z) = \sin^2\theta \,. \tag{4.4}$$

Since $\epsilon = 1 - \omega_{\mathrm{pe}}^2(z)/\omega^2$, reflection takes place where the electron plasma frequency $\omega_{\mathrm{pe}} = \omega\cos\theta$. An obliquely incident light wave reflects at a density lower than the critical density i.e., where $n_e = n_{cr} \cos^2 \theta$. Here the critical density is defined as $n_{cr} = m\omega^2/4\pi e^2$.

Our previous analyses can be carried out with the straightforward substitution $\epsilon(z) \rightarrow \epsilon(z) - \sin^2 \theta$. For example, in a plasma with a linear density profile, $n_e = n_{cr}z/L$, the wave reflects at $z = L\cos^2 \theta$, and the Airy function pattern occurs relative to this point rather than at $z = L$.

4.2   OBLIQUELY INCIDENT P-POLARIZED LIGHT WAVES — RESONANCE ABSORPTION

If the electric vector of the light wave lies in the plane of incidence, the light wave is termed p-polarized. In this case, there's a component of the electric vector which oscillates electrons along the direction of the density gradient i.e., $\mathbf{E} \cdot \nabla n_e \neq 0$. Since this oscillation generates fluctuations in charge density which can be resonantly enhanced by the plasma, the wave is no longer purely electromagnetic. Part of the energy of the incident light wave is transferred to an electrostatic oscillation (electron plasma wave), a phenomenon termed resonance absorption [1,2].

We again consider a plane electromagnetic wave incident with an angle θ onto an inhomogeneous plasma slab with density $n_e(z)$, as shown in Fig. 4.1. Now the electric vector is taken to be in the plane of incidence i.e., $\mathbf{E} = E_y \hat{y} + E_z \hat{z}$. It is readily seen that the field acquires an electrostatic component. Poisson's equation gives

$$\nabla \cdot (\epsilon \mathbf{E}) = 0,$$

where $\epsilon(z) = 1 - \omega_{pe}^2(z)/\omega^2$ is the dielectric function of the plasma. Since $\nabla \cdot (\epsilon \mathbf{E}) = \epsilon \nabla \cdot \mathbf{E} + \nabla \epsilon \cdot \mathbf{E}$, we obtain

$$\nabla \cdot \mathbf{E} = -\frac{1}{\epsilon} \frac{\partial \epsilon}{\partial z} E_z.$$

Note the resonant response when $\epsilon = 0$, i.e., where $\omega_{pe} = \omega$.

The physical interpretation is straightforward. Oscillation of electrons between regions of differing density directly creates a charge density fluctuation, δn, which is

$$\delta n = n_e(\mathbf{x} + \mathbf{x}_{os}) - n_e(\mathbf{x})$$
$$\simeq \mathbf{x}_{os} \cdot \nabla n_e,$$

where $\mathbf{x}_{os}$ is the amplitude of oscillation of an electron in the electric field of the light wave ($\mathbf{x}_{os} = e\mathbf{E}/m\omega^2$). Where $\omega = \omega_{pe}$, this imposed charge density fluctuation is at just the frequency at which the plasma resonantly responds. Hence an electron plasma wave is excited where $\epsilon = 0$; that is, at the critical density.

Even though an obliquely incident light wave reflects at a density less than the critical density, its fields still tunnel into the critical density region and excite the resonance. To determine the energy transfer to the excited plasma wave, we need to determine the size of the electric field along the density gradient near the critical density. In order to evaluate E_z, it is most convenient to work in terms of the magnetic field of the p-polarized wave. Noting that $\mathbf{B} = \hat{x}B_x$ and using the conservation of $k_y = (\omega/c)\sin\theta$, we express

$$\mathbf{B} = \hat{x}\, B(z)\, \exp\left(-i\,\omega t + \frac{i\,\omega\, y\,\sin\theta}{c}\right). \tag{4.5}$$

The electric field is related to the magnetic field by substituting Eq. (4.5) into Ampere's law:

$$\nabla \times \mathbf{B} = -\frac{i\,\omega}{c}\,\epsilon\,\mathbf{E}. \tag{4.6}$$

The z component of Eq. (4.6) then gives

$$E_z = \frac{\sin\theta\, B(z)}{\epsilon(z)}. \tag{4.7}$$

Since E_z is strongly peaked at the critical density, we approximate the resonantly driven field as $E_d/\epsilon(z)$, where E_d is evaluated at the resonance point. Physically, E_d is simply the component of the electric field of the light wave which oscillates electrons along the density gradient at the critical density i.e., the field driving the resonance.

To evaluate E_d, we need to calculate the magnetic field at the critical density. For our purposes, it suffices to simply estimate the value of the magnetic field using the insight obtained from our previous calculations of wave propagation in inhomogeneous plasmas. Assuming a linear density profile ($n_e = n_{cr}\, z/L$), we represent $B(z=L)$ as its value at the turning point $B(z=L\cos^2\theta)$ multiplied by an exponential decay from the turning point to the critical density. The value of B at the turning point is estimated using the Airy function solution for an s-polarized wave, which gives $B(z=L\cos^2\theta) \approx 0.9\, E_{FS}\, (c/\omega L)^{1/6}$. Here E_{FS} is the value of the electric field of the light wave in free space. The decay of the field as it penetrates beyond the turning point is estimated by $e^{-\beta}$, where

$$\beta = \int_{L\cos^2\theta}^{L} \frac{1}{c} \sqrt{\omega_{\text{pe}}^2 - \omega^2 \cos^2\theta} \; dz \; .$$

For a linear density profile, $\beta = (2\omega L/3c)\sin^3\theta$. Hence, we obtain

$$B(z=L) \simeq 0.9\, E_{\text{FS}} \left(\frac{c}{\omega L}\right)^{1/6} \exp\!\left(\frac{-2\omega L \sin^3\theta}{3c}\right) . \qquad (4.8)$$

The important physical features of resonance absorption can be deduced from our approximate treatment for E_d [3]. Using Eq. (4.8) and defining a new variable $\tau = (\omega L/c)^{1/3}\sin\theta$, we obtain

$$E_d = \frac{E_{\text{FS}}}{\sqrt{2\pi\,\omega L/c}}\, \phi(\tau) , \qquad (4.9)$$

where $\phi(\tau) \simeq 2.3\,\tau\,\exp(-2\tau^3/3)$. The driver field vanishes as $\tau \to 0$, since the component of the electric vector of the incident light wave along the density gradient varies as $\sin\theta$. Likewise, the driver field becomes very small for large τ, since the incident wave then has to tunnel through too large a distance to reach the critical density surface. Between these two limits, there is an optimium angle of incidence given approximately by $(\omega L/c)^{1/3}\sin\theta \simeq 0.8$.

In Fig. 4.2 we compare our simple estimate for $\phi(\tau)$ with the result obtained by Denisov [2] by numerically solving the wave equation. Note that our heuristic estimate is in reasonable agreement with the detailed calculation. As expected, our heuristic solution is quite accurate for $\tau \ll 1$, since our estimate for $B(z=L)$ becomes exact as $\tau \to 0$. Our expression for $\phi(\tau)$ is qualitatively correct even for $\tau \gg 1$, since the dominant physical effect is then the attenuation of the incident field as it tunnels from the cut-off density to the critical density.

Having related the electrostatic field near the critical density to the electric field and the angle of incidence of the light wave, we can now calculate the energy absorption. As shown in Eq. (4.7), the resonantly-driven field is $E_z = E_d/\epsilon(z)$. If we include a small damping of the wave with frequency ν, $\epsilon(z) = 1 - \omega_{\text{pe}}^2(z)/\omega(\omega + i\nu)$, as we will show in the next chapter. Hence, E_z has a resonance behavior near $z = L$, i.e., the maximum value of E_z is proportional to ν^{-1} and the width of the

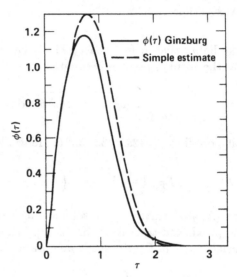

Figure 4.2   A plot of the function $\phi(\tau)$, which characterises the efficiency of resonance absorption. The solid line is from *Ginzberg* (1964).

resonant region is proportional to ν. This feature of E_z will enable us to compute the energy absorbed via excitation of the electrostatic wave without specifying the detailed value of ν (assuming $\nu/\omega \ll 1$). Hence, ν can represent dissipation by electron-ion collisions, linear or nonlinear wave-particle interactions, or even propagation of the wave out of the resonant region.

The absorbed energy flux (I_{abs}) is

$$I_{\text{abs}} = \int_0^\infty \nu \frac{E_z^2}{8\pi} \, dz = \frac{\nu}{8\pi} \int_0^\infty \frac{dz \, E_d^2(z)}{|\epsilon|^2} . \tag{4.10}$$

For a linear density profile ($n_e = n_{\text{cr}} z/L$), we have

$$|\epsilon|^2 = \left(1 - \frac{z}{L}\right)^2 + \left(\frac{\nu}{\omega}\right)^2 \frac{z^2}{L^2} . \tag{4.11}$$

Substituting Eq. (4.11) into Eq. (4.10) and approximating E_d as constant over the narrow width of the resonance function gives

$$I_{\text{abs}} \simeq \frac{\nu \, E_d^2(z=L)}{8\pi} \int_0^\infty \frac{dz}{\left(1 - z/L\right)^2 + \left(\nu/\omega\right)^2} . \tag{4.12}$$

The integral is easily evaluated to give $\pi\,\omega/\nu$. Hence

$$I_{\text{abs}} \simeq \frac{\omega\,L\,E_d^2}{8}\,. \tag{4.13}$$

By conservation of energy, $I_{\text{abs}} = f_A\,c\,E_{\text{FS}}^2/8\pi$, where f_A is the fractional absorption of the incident light wave due to the excitation of an electrostatic wave at the critical density. Substituting for $E_d(z=L)$ from Eq. (4.9), we then obtain $f_A \simeq \phi^2(\tau)/2$, where $\phi(\tau)$ is the characteristic resonance function describing the strength of the excitation as a function of the angle of incidence and the scale length of the density gradient.

Our simple model emphasizes the physics of resonance absorption and captures its basic features. For a linear density profile, the fractional absorption peaks at $\theta_{\text{max}} \approx \sin^{-1}[\,0.8(c/\omega L)^{1/3}\,]$ and is sizeable for a range of angles of incidence $\Delta\theta \sim \theta_{\text{max}}$. The peak absorption is somewhat overestimated. Detailed numerical calculations for a linear density profile show a peak resonance absorption of about 0.5 [4–6].

References

1. Ginzberg, V. L., *The Properties of Electromagnetic Waves in Plasma.* Pergamon, New York, 1964.

2. Denisov, N. G., On a singularity of the field of an electromagnetic wave propagated in an inhomogeneous plasma, *Sov. Phys.— JETP* 4, 544 (1957).

3. Kruer, W. L., Theory and simulation of laser plasma coupling; in *Laser Plasma Interactions*, (R. A. Cairns and J. J. Sanderson, eds.). SUSSP Publications, Edinburgh, 1980.

4. Forslund, D., J. Kindel, K. Lee, E. Lindman and R. L. Morse, Theory and simulation of resonant absorption in a hot plasma, *Phys. Rev. A* **11**, 679 (1975).

5. Estabrook, K. G., E. J. Valeo, and W. L. Kruer, Two-dimensional relativistic simulations of resonance absorption, *Phys. Fluids* **18**, 1151 (1975).

6. Speziale, T. and P. J. Catto, Linear wave conversion in an unmagnetized collisionless plasma, *Phys. Fluids* **20**, 990 (1977).

CHAPTER 5

Collisional Absorption of Electromagnetic Waves in Plasmas

So far we have focused on collisionless plasma behavior. As discussed in Chapter 1, collisional effects due to discrete particle encounters can be systematically neglected as the number of particles in a Debye sphere (N_D) becomes very large. Even when N_D is large, there are, of course, some collisions, which can be iteratively included by going to first order in the expansion parameter $1/N_D$. A collision term is then added to the Vlasov equation, which becomes

$$\frac{\partial f_j}{\partial t} + \mathbf{v}\cdot\frac{\partial f_j}{\partial \mathbf{x}} + \frac{q_j}{m_j}\left(\mathbf{E} + \frac{\mathbf{v}\times\mathbf{B}}{c}\right)\cdot\frac{\partial f_j}{\partial \mathbf{v}} = \sum_k \left(\frac{\partial f_{jk}}{\partial t}\right)_{\mathrm{C}},$$

where $(\partial f_{jk}/\partial t)_{\mathrm{C}}$ represents the rate of change of f_j due to collisions with the kth charge species.

The contribution of collisions to the moment equations is straightforward. If we neglect ionization and recombination, collisions do not change the number of charges of each species. Hence $\int d\mathbf{v}\ \sum_k (\partial f_{jk}/\partial t)_{\mathrm{C}} = 0$, and the continuity equation is unchanged. Noting that collisions between charges of the same species lead to no net change of momentum, we have

$$\int d\mathbf{v} \, \mathbf{v} \sum \left(\frac{\partial f_{jk}}{\partial t} \right)_C = \sum_{k \neq j} \left(\frac{\partial}{\partial t} n_j \mathbf{u}_j \right)_k ,$$

where $(\partial n_j \mathbf{u}_j / \partial t)_k$ denotes the rate of change in the momentum of the jth species due to collisions with charge species k. Hence the force equation becomes

$$n_j \left(\frac{\partial \mathbf{u}_j}{\partial t} + \mathbf{u}_j \cdot \frac{\partial}{\partial x} \mathbf{u}_j \right) =$$
$$n_j \frac{q_j}{m_j} \left(\mathbf{E} + \frac{\mathbf{u}_j \times \mathbf{B}}{c} \right) - \frac{1}{m_j} \frac{\partial p_j}{\partial \mathbf{x}} - \sum_{k \neq j} \left(\frac{\partial}{\partial t} n_j \mathbf{u}_j \right)_k . \tag{5.1}$$

5.1   COLLISIONAL DAMPING OF LIGHT WAVES

To investigate collisional damping, we again consider the linearized plasma response to a high-frequency field of the form $\mathbf{E}(\mathbf{x}) \, e^{-i\omega t}$. Treating the ions with charge state Z as a fixed, neutralizing background with density $n_e(x)/Z$, we need only treat the dynamics of the electron fluid. For an electron fluid with density n_e and velocity $\mathbf{u}_e$ interacting with a stationary ion fluid, it is convenient to express

$$\left(\frac{\partial}{\partial t} n_e \mathbf{u}_e \right)_i = \nu_{ei} \, n_e \mathbf{u}_e , \tag{5.2}$$

where ν_{ei} is a collision frequency which describes electron scattering by the ions.

This collision frequency ν_{ei} depends on an average over the velocity distribution of the electrons. Indeed, the form of the average depends on the physical process under consideration, and so a more detailed treatment is needed in order to derive the numerical value of ν_{ei}. Fortuitously, for a Maxwellian distribution of electron velocities, the electron-ion collision frequency which describes the damping of a high-frequency wave is essentially the same as the characteristic electron-ion collision frequency which we estimated in Eq. (1.2). A derivation of this result will be presented in Section 5.4.

The analysis is straightforward. If we use Eqs. (5.1) and (5.2), the linearized force equation for the electron fluid becomes

$$\frac{\partial \mathbf{u}_e}{\partial t} = - \frac{e}{m} \mathbf{E} - \nu_{ei} \mathbf{u}_e .$$ (5.3)

Since the field varies harmonically with time,

$$\mathbf{u}_e = \frac{-i\, e\, \mathbf{E}}{m(\omega + i\nu_{ei})} .$$

The plasma current density is then

$$\mathbf{J} = - n_e\, e\, \mathbf{u}_e = \frac{i\, \omega_{\mathrm{pe}}^2}{4\pi\,(\omega + i\nu_{ei})} \mathbf{E} ,$$

where ω_{pe} is the electron plasma frequency. Note that the plasma conductivity σ ($\mathbf{J} = \sigma \mathbf{E}$) is now complex: $\sigma = i\omega_{\mathrm{pe}}^2/[4\pi(\omega + i\nu_{ei})]$. Faraday's and Ampere's laws become

$$\nabla \times \mathbf{E} = \frac{i\omega}{c} \mathbf{B}$$ (5.4)

$$\nabla \times \mathbf{B} = \frac{4\pi}{c} \sigma \mathbf{E} - \frac{i\omega}{c} \mathbf{E} = - \frac{i\omega}{c} \epsilon \mathbf{E} ,$$ (5.5)

where the dielectric function of the plasma is now

$$\epsilon = 1 - \frac{\omega_{\mathrm{pe}}^2}{\omega(\omega + i\nu_{ei})} .$$ (5.6)

The wave equation for $\mathbf{E}$ is obtained by taking the curl of Eq. (5.4) and substituting for $\nabla \times \mathbf{B}$ from Eq. (5.5):

$$\nabla^2 \mathbf{E} - \nabla(\nabla \cdot \mathbf{E}) + \frac{\omega^2}{c^2} \epsilon \mathbf{E} = 0 .$$ (5.7)

Let us first derive the dispersion relation for light waves in a spatially uniform plasma. Taking $\mathbf{E}(x) \sim e^{i\mathbf{k} \cdot \mathbf{x}}$ and substituting for ϵ into Eq. (5.7) gives

$$\omega^2 = k^2 c^2 + \omega_{\mathrm{pe}}^2 \left(1 - \frac{i\nu_{ei}}{\omega} \right) ,$$ (5.8)

where we have assumed that $\nu_{ei}/\omega \ll 1$. The light waves are now damped. Expressing $\omega = \omega_r - i\nu/2$, where ν is the energy damping rate, we obtain

$$\omega_r = \left(\omega_{\text{pe}}^2 + k^2 c^2\right)^{1/2}$$

$$\nu = \frac{\omega_{\text{pe}}^2}{\omega_r^2}\,\nu_{ei}\,. \tag{5.9}$$

The collisional damping has a simple physical interpretation. The rate of energy loss from the light wave ($\nu E^2/8\pi$) must balance the rate at which the oscillatory energy of the electrons is randomized by the electron-ion scattering with a frequency ν_{ei} ($\nu_{ei}\,n_0\,m\,v_{\text{os}}^2/2$). Since $v_{\text{os}} = eE/m\omega_r$, this power balance gives $\nu = \nu_{ei}\,\omega_{\text{pe}}^2/\omega_r^2$.

It is instructive to also consider a spatial problem i.e., let ω be real and k be complex. Then, substituting $k = k_r + i\,k_i/2$ into Eq. (5.8) and assuming that $k_i \ll k_r$, we obtain

$$k_r = \frac{1}{c}\,\sqrt{\omega^2 - \omega_{\text{pe}}^2}$$

$$k_i = \frac{\omega_{\text{pe}}^2}{\omega^2}\,\frac{\nu_{ei}}{v_g}\,,$$

where k_i is rate at which the energy decays in space and v_g is the group velocity of the light wave. Note that the energy damping length (k_i^{-1}) is simply v_g/ν, where ν is given in Eq. (5.9).

5.2 COLLISIONAL DAMPING OF A LIGHT WAVE IN AN INHOMOGENEOUS PLASMA

Let us now investigate collisional damping of a light wave propagating into an inhomogeneous plasma. First, we will neglect the density dependence of the collision frequency and compute the absorption of a normally incident wave both from an analytic solution for a plasma with a linear density profile and from a WKB treatment. Then we will use the WKB treatment to allow for the density dependence of the collision frequency and for oblique incidence.

We start by again considering a plane wave propagating in the z direction into a plasma slab with electron density $n_e(z)$. Since there is only variation in the z direction, Eq. (5.7) becomes

$$\frac{\mathrm{d}^2 E}{\mathrm{d}z^2} + \frac{\omega^2}{c^2}\,\epsilon(z)\,E = 0\,, \tag{5.10}$$

where we have taken $\mathbf{E} = E\,\hat{x}$. If we assume a linear density profile ($n_e = n_{cr}\,z/L$) and neglect the dependence of ν_{ei} on the plasma density,

$$\epsilon = 1 - \frac{z}{L\left(1 + i\nu_{ei}^*/\omega\right)} \quad , \tag{5.11}$$

where ν_{ei} is approximated by its value (ν_{ei}^*) at the critical density. This is a reasonable first approximation, since most of the collisional absorption occurs near the critical density where the electron-ion collision frequency maximizes and the group velocity of the light wave minimizes. Note that we are also assuming that the plasma with a density less than or equal to the critical density is isothermal. Substituting Eq. (5.11) into Eq. (5.10) gives

$$\frac{d^2 E}{dz^2} + \frac{\omega^2}{c^2}\left[1 - \frac{z}{L\left(1 + \frac{i\nu_{ei}^*}{\omega}\right)}\right] E = 0\,.$$

By changing variables, we again obtain Airy's equation

$$\frac{d^2 E}{d\eta^2} - \eta E = 0\,, \tag{5.12}$$

where η is now a complex variable:

$$\eta = \left[\frac{\omega^2}{c^2 L\left(1 + \frac{i\nu_{ei}^*}{\omega}\right)}\right]^{1/3} \left[z - L\left(1 + \frac{i\nu_{ei}^*}{\omega}\right)\right]\,. \tag{5.13}$$

As discussed in Chapter 3, the solution which satisfies physically reasonable boundary conditions is

$$E(\eta) = \alpha\,A_i(\eta)\,, \tag{5.14}$$

where $A_i(\eta)$ is a well-documented Airy function. The constant α is chosen by matching to the incident light wave at the vacuum-plasma interface at $z = 0$. For $|\eta| \gg 1$, we can evaluate $E(\eta)$ using the asymptotic representation for $A_i(\eta)$ i.e.,

$$A_i(-\eta) = \frac{1}{\sqrt{\pi}\,\eta^{1/4}}\,\cos\left(\frac{2}{3}\eta^{3/2} - \frac{\pi}{4}\right)\,.$$

Hence at $z = 0$, E can be represented as an incident plus a reflected wave whose amplitude is multiplied by the quantity $e^{i\phi}$, where

$$\phi = \frac{4}{3}\left[-\eta(z=0)\right]^{3/2} - \frac{\pi}{2}.$$

Since η is now complex, there is both a phase shift and a damping of the reflected wave. At $z = 0$, $\eta = -\left[(\omega L/c)(1 + i\nu_{ei}^*/\omega)\right]^{2/3}$. For $\nu_{ei}^*/\omega \ll 1$, the phase shift (the real part of ϕ) is the same as in the collisionless calculation i.e., $\phi_{\text{real}} = (4\omega L/3c) - \pi/2$. The imaginary part of ϕ is $4\nu_{ei}^* L/3c$, which means that the reflected wave is decreased in amplitude by $\exp(-4\nu_{ei}^* L/3c)$, or in energy by $\exp(-8\nu_{ei}^* L/3c)$. Hence the fractional absorption f_A due to collisional damping is

$$f_A = 1 - \exp\left(-\frac{8\nu_{ei}^* L}{3c}\right). \tag{5.15}$$

Let's now calculate the collisional absorption in a plasma with a linear density profile using WKB theory. Here

$$E \sim \exp\left[i\int^z k(z')\,dz'\right],$$

where

$$k(z') = \frac{\omega}{c}\left[\epsilon(z')\right]^{1/2}.$$

The energy of the wave decreases by the factor $e^{-2\delta}$, where

$$\delta = 2\frac{\upsilon}{c}\int_0^L \Im\left(\epsilon^{1/2}\right)\,dz'.$$

The symbol $\Im$ denotes the imaginary part, and the factor of two enters since the wave is absorbed as it propagates both into and out of the plasma. Substituting for ϵ from Eq. (5.11) gives

$$\delta = \frac{2\omega}{c}\Im\int_0^L\left[1 - \frac{z'}{L\left(1 + \frac{i\nu_{ei}^*}{\omega}\right)}\right]^{1/2}\,dz'.$$

For $\nu_{ei}^*/\omega \ll 1$, $\delta = 4\nu_{ei}^* L/3c$. Hence the energy of the light wave decays by the factor $\exp(-8\nu_{ei}^* L/3c)$, the same result as that given by the analytic solution.

5.3   COLLISIONAL ABSORPTION INCLUDING OBLIQUE INCIDENCE AND A DENSITY DEPENDENT COLLISION FREQUENCY

Using WKB theory, let's now extend our calculations of collisional absorption to include obliquely-incident light waves and to allow for the dependence of the collision frequency on density. For definiteness, we consider s-polarized plane waves incident onto a plasma slab with electron density $n_e(z)$. As discussed in Chapter 4, the local dispersion relation then is

$$\frac{k_z^2 c^2}{\omega^2} = \epsilon(z) - \sin^2 \theta , \qquad (5.16)$$

where θ is the angle of incidence. Substituting Eq. (5.11) into Eq. (5.16) gives

$$k_z = \frac{\omega}{c} \left[\cos^2 \theta - \frac{\omega_{pe}^2}{\omega (\omega + i \nu_{ei})} \right]^{1/2} . \qquad (5.17)$$

The density dependence of ν_{ei} is easily included. Referring to Eq. (1.2) of Chapter 1, we neglect any weak dependence on density introduced by $\ln \Lambda$ and note that the collision frequency is then simply proportional to the plasma density. Hence we will approximate $\nu_{ei} = \nu_{ei}^* \, n_e/n_{cr}$, where ν_{ei}^* is again the collision frequency evaluated at the critical density.

In the WKB approximation, the wave energy decays by $e^{-2\delta}$, where

$$\delta = 2 \, \Im m \int_0^{L_t} k_z(z') \, dz' ,$$

and L_t is the turning point. If we assume a linear density profile and use Eq. (5.17), we obtain

$$\delta = \frac{2\omega}{c} \, \Im m \int_0^{L \cos^2 \theta} \left[\cos^2 \theta - \frac{z'}{L} \left(1 - \frac{i \nu_{ei}^*}{\omega} \frac{z'}{L} \right) \right]^{1/2} dz' ,$$

where we have assumed that $\nu_{ei}^*/\omega \ll 1$. This standard integral gives $\delta = (16\nu_{ei}^* L/15c) \cos^5 \theta$. Hence the fractional absorption is

$$f_A = 1 - \exp\left(-\frac{32 \, \nu_{ei}^* \, L}{15c} \cos^5 \theta \right) . \qquad (5.18)$$

For normal incidence ($\theta = 0$), the density-dependence of ν_{ei} has reduced the coefficient in the exponent of Eq. (5.15) from 8/3 to 32/15 i.e., by about 20%. Note also that the absorption is a sensitive function of the angle of incidence. Since an obliquely incident wave reflects at a lower density, less collisional plasma is traversed. Finally, we note that the collisional absorption will depend in detail on the density profile of the plasma. For example, for an exponential profile ($n_e = n_{cr} \exp(-z/L)$),

$$f_A = 1 - \exp\left(- \frac{8\,\nu_{ei}^* L}{3c} \cos^3 \theta\right).$$

5.4   DERIVATION OF THE DAMPING COEFFICIENT

As a final topic, let us give a derivation of the collisional damping rate of a light wave. To determine the electron-ion collision frequency heuristically introduced in Eq. (5.2), we need to start with a representation for $(\partial f_{ei}/\partial t)_C$. The simplest description starts with the Fokker-Planck equation

$$\left(\frac{\partial f_{ei}}{\partial t}\right)_C = - \frac{\partial}{\partial \mathbf{v}} \cdot \left(f_e \langle \Delta \mathbf{v} \rangle\right) + \frac{1}{2} \frac{\partial^2}{\partial \mathbf{v}\,\partial \mathbf{v}} : \left(f_e \langle \Delta \mathbf{v}\, \Delta \mathbf{v} \rangle\right),$$

where $\langle \Delta \mathbf{v} \rangle$ describes the slowing down of electrons due to electron-ion encounters and $\langle \Delta \mathbf{v} \Delta \mathbf{v} \rangle$ their diffusion in velocity space. These coefficients can be derived by computing the changes in velocity of electrons streaming past ions and summing over encounters i.e., by a more detailed treatment of the approach used in Chapter 1 to estimate the 90° collision frequency. With this motivation, we will simply give the standard result,

$$\left(\frac{\partial f_{ei}}{\partial t}\right)_C = A \frac{\partial}{\partial \mathbf{v}} \cdot \left[\frac{v^2 \underset{\approx}{\mathbf{I}} - \mathbf{v}\,\mathbf{v}}{v^3} \cdot \frac{\partial f_e}{\partial \mathbf{v}}\right]. \qquad (5.19)$$

Here $A = (2\pi\, n_e\, Z\, e^4/m^2)\ln\Lambda$ where n_e is the electron density, Z is the ion charge state and Λ is the ratio of the maximum and minimum impact parameters as discussed in Chapter 1. A detailed discussion of the derivation of Eq. (5.19) is given in Chapter 7 of *Shkarofsky et al.*, 1966.

To calculate the high-frequency resistivity, we consider a plasma with a uniform electron density n_e and a fixed, neutralizing background of ions. The electric field of the light wave is treated in the dipole approximation

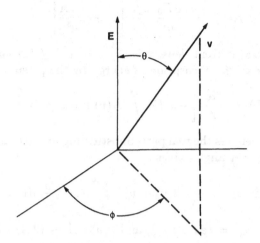

Figure 5.1   A sketch of the coordinate system.

i.e., $\mathbf{E} = \mathbf{E}_0 \cos \omega t$. If we substitute Eq. (5.19) into Eq. (5.1), the kinetic equation for the electrons becomes

$$\frac{\partial f_e}{\partial t} - \frac{e}{m} \mathbf{E} \cdot \frac{\partial f_e}{\partial \mathbf{v}} = A \frac{\partial}{\partial \mathbf{v}} \cdot \left[\frac{v^2 \underset{\approx}{\mathbf{I}} - \mathbf{v}\mathbf{v}}{v^3} \cdot \frac{\partial f_e}{\partial \mathbf{v}} \right] + C_{ee}(f_e) , \quad (5.20)$$

where f_e is the electron distribution function and $C_{ee}(f_e)$ denotes a similar but more complex operator describing electron-electron collisions. These electron-electron collisions are important for determining the form of the zero-order distribution function but can be neglected otherwise.

For low-intensity light, we decompose the distribution function into a zero-order part which depends only on the absolute value of $\mathbf{v}$ plus a perturbation driven by the field i.e., $f(\mathbf{v}) = f_0(v) + f_1(v) \cos \theta$, where θ is the angle between $\mathbf{v}$ and $\mathbf{E}$ as shown in Fig. 5.1. The linearized kinetic equation then becomes

$$\frac{\partial f_1}{\partial t} - \frac{e E}{m} \frac{\partial f_0}{\partial v} = - \frac{2 A}{v^3} f_1 . \quad (5.21)$$

Here we have used the fact that the collision operator vanishes for any function of $|\mathbf{v}| = v$ and

$$A \frac{\partial}{\partial \mathbf{v}} \cdot \frac{v^2 \underset{\approx}{\mathbf{I}} - \mathbf{v}\mathbf{v}}{v^3} \cdot \frac{\partial}{\partial \mathbf{v}} f_1(v) \cos \theta = - \frac{2 A}{v^3} f_1(v) \cos \theta .$$

Since $\mathbf{E} = \mathbf{E}_0\, e^{-i\omega t}$, the driven solution of Eq. (5.21) is

$$f_1(v) = \frac{i\, e\, E_0}{m} \frac{\partial f_0}{\partial v} \left[\omega + \frac{i\, 2\, A}{v^3} \right]^{-1} . \qquad (5.22)$$

The perturbed current density is $\mathbf{J}_1 = -e \int f_1(v)\, \cos\theta\, \mathbf{v}\, d\mathbf{v}$, and the average rate of the absorption of energy by the plasma is

$$\langle\, \mathbf{J}_1 \cdot \mathbf{E}\, \rangle = \frac{\Re e}{2} \left[-e\, E_0 \int f_1(v)\, v\, \cos^2\theta\, d\mathbf{v} \right] . \qquad (5.23)$$

The symbol $\Re e$ denotes the real part. Substituting for $f_1(v)$ from Eq. (5.22) and integrating over angles gives

$$\langle\, \mathbf{J}_1 \cdot \mathbf{E}\, \rangle = -\frac{4\,\pi}{3}\, A\, n_e\, m\, v_{\text{os}}^2 \int_0^\infty dv\, \frac{\partial \bar{f}_0}{\partial v}\, g(v) . \qquad (5.24)$$

Here $f_0 = n_e \bar{f}_0$, $v_{\text{os}} = (e\, E_0/m\omega)$, and $g(v) = [1 + (2\, A/v^3\omega)^2\,]^{-1}$. Note that $(2\, A/v^3\omega) \sim \nu_{ei}(v)/\omega$, where $\nu_{ei}(v)$ is a characteristic collision frequency for electrons with velocity v. Since $\nu_{ei}(v)/\omega \ll 1$ for all but a small class of electrons, we approximate $g(v) \approx 1$ in the integral, giving

$$\int_0^\infty dv\, \frac{\partial f_0}{\partial v}\, g(v) = -\bar{f}_0(0) .$$

Finally, we invoke energy balance to equate $\langle \mathbf{J}_1 \cdot \mathbf{E} \rangle$ with the rate of energy loss from the field, which is $\nu\, E^2/8\,\pi$. If we use Eq. (5.24), the rate ν at which energy is damped by electron-ion collisions becomes

$$\nu = \frac{\omega_{\text{pe}}^2}{\omega^2}\, \frac{8\,\pi}{3}\, A\, \bar{f}_0(0) , \qquad (5.25)$$

where ω_{pe} is the electron plasma frequency.

The damping rate depends on the zero-order distribution function. The form of this distribution function in turn depends on whether electron-electron collisions (with frequency ν_{ee}) can equilibrate the distribution faster than electron-ion collisions cause it to heat. If $\nu_{ee}\, v_e^2 \gg \nu_{ei}\, v_{\text{os}}^2$ (i.e., if $(Z\, v_{\text{os}}^2/v_e^2) \ll 1$), the distribution function remains Maxwellian. Evaluation of Eq. (5.25) for a Maxwellian distribution then gives the result usually quoted in the literature

$$\nu = \frac{\omega_{\text{pe}}^2}{\omega^2}\, \frac{1}{3(2\,\pi)^{3/2}}\, \frac{Z\, \omega_{\text{pe}}^4}{n\, v_e^3}\, \ln \Lambda , \qquad (5.26)$$

where v_e is the electron thermal velocity. A modification of the usual expression for $\ln \Lambda$ should be noted. The maximum impact parameter is now v_e/ω rather than the electron Debye length. Since $\nu = \nu_{ei}\, \omega_{pe}^2/\omega^2$, Eq. (5.26) determines the electron-ion collision frequency which describes the damping of a high frequency wave in a plasma. If the density is expressed in cm^{-3} and the electron temperature in ev,

$$\nu_{ei} \simeq 3 \times 10^{-6}\, \ln \Lambda\, \frac{n_e\, Z}{\theta_{ev}^{3/2}}\,. \tag{5.27}$$

However, if $Z(v_{os}/v_e)^2 \gg 1$, electron-electron collisions cannot equilibrate the distribution function sufficiently rapidly. The form of the distribution function becomes determined by the collisional heating [4,5]. In this limit, we return to Eq. (5.20) and balance $\partial f_0/\partial t$ with the heating term,

$$\frac{\partial f_0}{\partial t} = \left\langle \frac{eE}{m} \cdot \frac{\partial f_1}{\partial v} \right\rangle,$$

where the brackets denote an average over angles. Noting that $f_1 = f_1(v)\cos\theta$ and averaging over angles gives

$$\frac{\partial f_0}{\partial t} = \frac{eE}{m}\, \frac{1}{3v^2}\, \frac{\partial}{\partial v}\left(v^2\, f_1(v)\right). \tag{5.28}$$

Substituting for $f_1(v)$ from Eq. (5.22), approximating $g(v) \approx 1$, and looking for a self-similar solution, we find

$$f_0 \sim \frac{1}{u^3}\, \exp\left[-\frac{1}{5}\left(\frac{v}{u}\right)^5\right], \tag{5.29}$$

where

$$u = \left(\frac{5\,A\,v_{os}^2}{3}\,t\right)^{1/5}.$$

Hence the self-consistent distribution function is super-Gaussian in this limit. Since this distribution has fewer particles than a Maxwellian near $v = 0$, the collisional damping rate is reduced by a factor of about 2.

References

1. Shkarofsky, I. P., T. W. Johnston, and M. P. Backynski, *The Particle Kinetics of Plasmas.* Addison-Wesley, Mass., 1966.

2. Dawson, J. M. and C. Oberman, High frequency conductivity and the emission and absorption coefficients of a fully ionized plasma, *Phys. Fluids* **5**, 517 (1962).

3. Johnston, T. W. and J. M. Dawson, Correct values for high-frequency power absorption by inverse bremsstrahlung in plasmas, *Phys. Fluids* **16**, 722 (1973).

4. Langdon, A. B., Nonlinear inverse bremsstrahlung and heated-electron distributions, *Phys. Rev. Letters* **44**, 575 (1980).

5. Jones, R. D. and K. Lee, Kinetic theory, transport, and hydrodynamics of a high-Z plasma in the presence of an intense field, *Phys. Fluids* **25**, 2307 (1982).

CHAPTER 6

Parametric Excitation of Electron and Ion Waves

As we have seen in our consideration of resonance absorption, the oscillation of electrons in the direction of a spatial variation in the plasma density drives charge density fluctuations. When the frequency of the oscillation is near the electron plasma frequency, an electron plasma wave is resonantly excited. In the case of resonance absorption, the spatial variation was due to the density gradient produced by plasma expansion into a vacuum. However, the spatial variation in the density can also be due to ion density fluctuations associated with ion waves.

We will first discuss this coupling of a light wave into an electron plasma wave by an ion density fluctuation. Then we will show that this coupling can lead to unstable growth of both electron and ion waves, when account is taken of the generation of ion density fluctuations via the so-called ponderomotive force. A physical picture of the instability will be given, followed by a derivation from the two-fluid model.

6.1   COUPLING VIA ION DENSITY FLUCTUATIONS

To investigate the coupling of a light wave into an electron plasma wave by an ion density fluctuation, a simple, one-dimensional model is sufficient. First, we model the light wave as a spatially homogeneous oscillating electric field: $\mathbf{E}_0 = \hat{x} E_0 \exp(-i\omega t)$. In other words, the wave number of the light wave is neglected on the assumption that it is much less than k, the wave number of the fluctuation in ion density. Since the frequency of an ion wave is much less than the frequency of a light wave, we describe the ion density fluctuation as a static modulation in the plasma density, $n = n_0 + \Delta n \cos kx$, where n_0 is the average density and Δn is the amplitude of the density fluctuation. Finally, we treat the ions as fixed on this time scale and describe the electrons as a fluid with density n_e, mean velocity u_e, and pressure p_e.

To derive an equation for the high frequency electron density fluctuations, we take a time derivative of the continuity equation, a spatial derivative of the force equation, and combine to obtain

$$\frac{\partial^2 n_e}{\partial t^2} - \frac{\partial^2 n_e u_e^2}{\partial x^2} - \frac{e}{m}\frac{\partial n_e E}{\partial x} - \frac{1}{m}\frac{\partial^2 p_e}{\partial x^2} - \nu_{ei}\frac{\partial n_e u_e}{\partial x} = 0 , \qquad (6.1)$$

where we have included collisions with frequency ν_{ei}. We next linearize this equation, i.e., let $n_e = n_0 + \Delta n \cos kx + \tilde{n}$, $E = E_0 + \tilde{E}$ and $u_e = u_0 + \tilde{u}$, where the tilde denotes a small perturbation and u_0 is the oscillation velocity of electrons in the field E_0. If we treat $\tilde{n} \ll \Delta n \ll n_0$, Eq. (6.1) becomes

$$\frac{\partial^2 \tilde{n}}{\partial t^2} - \frac{e}{m}n_0\frac{\partial \tilde{E}}{\partial x} - 3v_e^2\frac{\partial^2 \tilde{n}}{\partial x^2} + \nu_{ei}\frac{\partial \tilde{n}}{\partial t} = -\frac{e E_0}{m}\Delta n\, k \sin kx . \qquad (6.2)$$

We have used an adiabatic equation of state, assuming $\omega/k \gg v_e$, the electron thermal velocity. Then by substituting the Poisson equation $(\partial \tilde{E}/\partial x = -4\pi e\tilde{n})$ into Eq. (6.2), we obtain

$$\frac{\partial}{\partial x}\left[\frac{\partial^2 \tilde{E}}{\partial t^2} + \omega_{pe}^2\, \tilde{E} - 3v_e^2\frac{\partial^2 \tilde{E}}{\partial x^2} + \nu_{ei}\frac{\partial \tilde{E}}{\partial t} \right] = \frac{4\pi e^2}{m}E_0\,\Delta n\, k\, \sin kx ,$$

where $\omega_{pe}^2 = 4\pi n_0 e^2/m$. Integration gives

$$\left(\frac{\partial^2}{\partial t^2} + \nu_{ei}\frac{\partial}{\partial t} + \omega_{pe}^2 - 3v_e^2\frac{\partial^2}{\partial x^2} \right)\tilde{E} = -\omega_{pe}^2\frac{\Delta n}{n_0}E_0 \cos kx . \qquad (6.3)$$

Equation (6.3) describes the excitation of an electron plasma wave by interaction of the pump field (light wave) with an ion density fluctuation.

The driven solution to Eq. (6.3) is straightforward. Noting that the pump field varies as $\exp(-i\omega t)$, we obtain

$$E = \frac{\omega_{pe}^2}{\omega^2} \frac{\Delta n}{n_0} \frac{E_0 \cos kx}{\epsilon(k,\omega)} \,, \tag{6.4}$$

where

$$\epsilon(k,\omega) = 1 - \frac{\omega_{pe}^2 + 3k^2 v_e^2}{\omega^2} + \frac{i\,\nu_{ei}}{\omega}\,.$$

Since energy is coupled into the driven wave, the pump field is damped. The energy damping rate, ν^*, is determined by balancing the rate of energy lost ($\nu^* E_0^2/8\pi$) with the rate of energy absorption via the driven wave i.e.,

$$\frac{\nu^* E_0^2}{8\pi} = \frac{k}{2\pi} \int_0^{2\pi/k} dx \; \frac{\nu_{ei}\,|E|^2}{8\pi}\,.$$

By substituting for E from Eq. (6.4) and noting that $\Im\mathrm{m}\,\epsilon = \nu_{ei}/\omega$, we then obtain [1,2]

$$\frac{\nu^*}{\omega} = \frac{1}{2}\left(\frac{\Delta n}{n_{cr}}\right)^2 \frac{\Im\mathrm{m}\,\epsilon}{|\epsilon(k,\omega)|^2}\,.$$

Here n_{cr} is the critical density determined by the condition $\omega_{pe}^2 = \omega^2$.

We note that the electric field becomes very large when $\epsilon(k,\omega) \simeq 0$ i.e., when the pump field resonantly couples to an electron plasma wave. Of course, our linearized analysis fails if $\epsilon(k,\omega)$ becomes too small, and nonlinear effects must then be considered. One such nonlinear effect is repeated mode coupling. When $|\Delta n/n_{cr}| > |\epsilon(k,\omega)|$, the driven wave becomes as large as the pump. It in turn acts like a pump to drive a wave at $2k$, which in turn can beat with the ion density fluctuation to drive a plasma wave at $3k$, and so on. A spectrum of driven waves is obtained; the maximum wave number can be estimated by the condition $|\epsilon(Nk,\omega)| \sim \Delta n/n_{cr}$. Note that even a modest density fluctuation can efficiently couple a long wavelength plasma wave into shorter wavelength ones.

6.2   THE PONDEROMOTIVE FORCE

The coupling of a light wave into electron plasma waves by density fluctuations is a very basic phenomenon, which emphasizes that electromagnetic and electrostatic waves are inherently coupled in a turbulent plasma. If sizeable levels of ion fluctuations exist in the plasma, this coupling can clearly be very significant. In fact, sizeable ion density fluctuations can be self-consistently produced in the light-plasma interaction, since an excited plasma wave beats with the light wave to generate variations in electric field pressure. This gradient in field pressure gives rise to a force (the so-called ponderomotive force), which acts to generate ion density fluctuations.

To introduce the ponderomotive force, we consider the response of a homogeneous plasma to a high frequency field whose amplitude is spatially dependent i.e., $\mathbf{E} = \mathbf{E}(\mathbf{x}) \sin \omega t$, where $\omega \gtrsim \omega_{\mathrm{pe}} \gg \omega_{\mathrm{pi}}$. We treat the electrons as a fluid and compute their response to order E^2. If we neglect the electron pressure, the force equation is

$$\frac{\partial \mathbf{u}_e}{\partial t} + \mathbf{u}_e \cdot \nabla \mathbf{u}_e = -\frac{e}{m} \mathbf{E}(\mathbf{x}) \sin \omega t . \qquad (6.6)$$

To lowest order in $|\mathbf{E}|$, $\mathbf{u}_e = \mathbf{u}^h$ where

$$\begin{aligned} \frac{\partial \mathbf{u}^h}{\partial t} &= -\frac{e}{m} \mathbf{E}(\mathbf{x}) \sin \omega t , \\ \mathbf{u}^h &= \frac{e\,\mathbf{E}}{m\omega} \cos \omega t . \end{aligned} \qquad (6.7)$$

The electrons are simply oscillating in the local electric field. By averaging the force equation over these rapid oscillations, we obtain

$$m \frac{\partial \mathbf{u}^s}{\partial t} = -e\,\mathbf{E}^s - m \langle \mathbf{u}^h \cdot \nabla \mathbf{u}^h \rangle_t , \qquad (6.8)$$

where $\langle \ \rangle_t$ denotes an average over high frequency oscillation and $\mathbf{u}^s = \langle u_e \rangle_t$, $\mathbf{E}^s = \langle \mathbf{E} \rangle_t$. Substituting for $\mathbf{u}^h$ from Eq. (6.7), we obtain

$$m \frac{\partial \mathbf{u}^s}{\partial t} = -e\,\mathbf{E}^s - \frac{1}{4} \frac{e^2}{m\omega^2} \nabla \mathbf{E}^2(\mathbf{x}) . \qquad (6.9)$$

Observe that the electrons experience a force which pushes them away from regions of high field pressure. This ponderomotive force ($\mathbf{F}_p$) is proportional to the gradient of the electric field pressure i.e.,

$$\mathbf{F}_p = -\frac{e^2}{4m\omega^2}\,\nabla\mathbf{E}^2(\mathbf{x})\,.$$

In a uniform plasma with density n, the ponderomotive force density $\mathbf{f}_p$ can be expressed as $\mathbf{f}_p = -\nabla(nm\langle u_h^2\rangle/2)$. In other words, the time-averaged energy density of motion in the electric field plays the same role as the ordinary pressure, which represents the random or thermal energy density.

6.3   INSTABILITIES — A PHYSICAL PICTURE

We can now easily see that a light wave can excite an instability in which both ion waves and electron plasma waves grow. An ion density fluctuation couples the light wave into an electron plasma wave. In turn, the electron plasma wave beats with the light wave to generate a spatial variation in the electric field intensity, which can enhance the ion density fluctuation via the ponderomotive force. Hence, there's a feed-back loop, and instability can result.

A physical picture of the instability can be given [3]. Consider first a static ion density fluctuation in an otherwise homogeneous plasma i.e., $n = n_0 + \Delta n \cos kx$. The electric field of the light wave is again approximated as a spatially homogeneous field of the form $E_d = E_0 \sin \omega t$. The electrostatic field E associated with the excited plasma wave is then given by Eq. (6.4). Explicitly including the time dependence and neglecting collisions, we then have

$$E = \frac{\omega_{pe}^2}{\omega^2 - \omega_{ek}^2}\,\frac{\Delta n}{n_0}\,E_0\,\cos kx\,\sin \omega t\,,$$

where $\omega_{ek}^2 = \omega_{pe}^2 + 3k^2 v_e^2$. Since E has a spatial dependence, the time-averaged electric field intensity has a gradient. To lowest order in the small amplitude Δn of the (thermal) ion density fluctuation, we obtain

$$\nabla\left\langle (E + E_d)^2 \right\rangle = -\frac{\omega_{pe}^2}{\omega^2 - \omega_{ek}^2}\,\frac{\Delta n}{n_0}\,E_0^2\,k\,\sin kx\,.$$

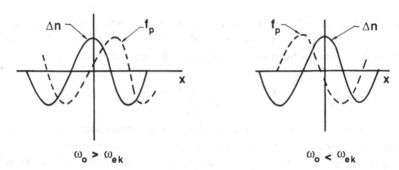

Figure 6.1   When $\omega_0 < \omega_{ek}$, the ponderomotive force acts to enhance the density fluctuation, i.e., to push more plasma into regions of higher density.

The ponderomotive force F_p is then

$$F_p = \frac{e^2 E_0^2}{2\,m\,\omega^2} \frac{\omega_{pe}^2}{\omega^2 - \omega_{ek}^2} \frac{\Delta n}{n_0} k \sin kx .$$

As shown in Fig. 6.1, the ponderomotive force acts to reduce the density fluctuation when $\omega > \omega_{ek}$. However, when $\omega < \omega_{ek}$, the ponderomotive force acts to *enhance* the density fluctuation. Hence a purely growing (zero frequency) ion density fluctuation will spontaneously grow from the noise. As it grows in amplitude, so also does the associated electron plasma wave. This instability is called the oscillating two stream.

We next consider the ion density fluctuation associated with an ion acoustic wave. The fluctuation is no longer static, but has a frequency equal to kv_s, where v_s is the ion sound velocity. In this case, the instability is most easily thought of as the resonant decay of the light wave into an electron plasma wave plus an ion acoustic wave. The instability is strongest when all three waves are matched in frequency i.e., when $\omega = \omega_{ek} + kv_s$. Hence this instability is often called the ion acoustic decay.

6.4   INSTABILITY ANALYSIS

We can derive these instabilities [4–9] from the two-fluid plasma description. For simplicity, we consider a spatially uniform plasma driven by a pump field of the form $\mathbf{E}_d = \mathbf{E}_0 \cos \omega_0 t$, where ω_0 is near ω_{pe}, the electron plasma frequency. We further consider only electrostatic perturbations and restrict the analysis to one-dimensional perturbations along the direction of the pump field.

The fluid equations for the electrons are

$$\frac{\partial n_e}{\partial t} + \frac{\partial n_e u_e}{\partial x} = 0 \qquad (6.11)$$

$$\frac{\partial u_e}{\partial t} + u_e \frac{\partial u_e}{\partial x} = -\frac{e}{m} E - \frac{1}{m n_e} \frac{\partial p_e}{\partial x} - \nu_e u_e \qquad (6.12)$$

$$\frac{p_e}{n_e^\gamma} = \text{constant} , \qquad (6.13)$$

where $\gamma = 3$ for high frequency perturbations and $\gamma = 1$ for low frequency ones. Note that we have included collisions with frequency ν_e to model either collisional or Landau damping of the electron waves. To proceed, we divide the electron fluctuations into low and high frequency components

$$\begin{aligned} n_e &= n_0 + n_e^\ell + n_e^h \\ u_e &= v_{os} + u_e^\ell + u_e^h , \end{aligned} \qquad (6.14)$$

where the superscripts ℓ and h denote low and high frequency, respectively, n_0 is the uniform background density, and v_{os} is the oscillation velocity of electrons in the pump field. To analyze for instability, we linearize by assuming that n_e^ℓ or $n_e^h \ll n_0$ and u_e^ℓ or $u_e^h \ll v_{os}$ and neglecting products of the perturbed quantities.

For spatially dependent electrostatic fluctuations, we can write

$$\frac{\partial E}{\partial t} + 4\pi J = 0 , \qquad (6.15)$$

where J is the longitudinal part of the current density. Equation (6.15) is readily obtained from Poisson's equation and the equation for continuity of charge. Linearizing and taking the high frequency component gives

$$\frac{\partial E^h}{\partial t} = 4\pi e \left(n_0 u_e^h + n_e^\ell v_{os} \right) . \qquad (6.16)$$

In turn, the high frequency component of the linearized force equation becomes

$$\frac{\partial u_e^h}{\partial t} + v_{os} \frac{\partial u_e^\ell}{\partial x} = -\frac{e}{m} E^h - \frac{3 v_e^2}{n_0} \frac{\partial n_e^h}{\partial x} - \nu_e u_e^h , \qquad (6.17)$$

where we have used $p_e^h = 3mv_e^2 n_e^h$. Taking a time derivative of Eq. 6.17 and substituting from Eq. (6.16) gives

$$\frac{\partial^2 u_e^h}{\partial t^2} + \nu_e \frac{\partial u_e^h}{\partial t} + \frac{\partial}{\partial t}\left(v_{os}\frac{\partial u_e^\ell}{\partial x}\right) + \frac{3v_e^2}{n_0}\frac{\partial^2 n_e^h}{\partial t \partial x} =$$
$$-\omega_{pe}^2 u_e^h - \frac{4\pi e^2}{m} n_e^\ell v_{os} . \quad (6.18)$$

The third term in Eq. (6.18) can be neglected relative to the other terms. First note that

$$\frac{\partial}{\partial t}\left(v_{os}\frac{\partial u_e^\ell}{\partial x}\right) \simeq \frac{\partial v_{os}}{\partial t}\frac{\partial u_e^\ell}{\partial x} ,$$

since the low frequency is assumed to be $\ll \omega_0 \sim \omega_{pe}$. We next use the low frequency component of the continuity equation to give

$$\frac{\partial v_{os}}{\partial t}\frac{\partial u_e^\ell}{\partial x} = -\frac{\partial v_{os}}{\partial t}\frac{1}{n_0}\left(\frac{\partial n_e^\ell}{\partial t} + v_{os}\frac{\partial n_e^h}{\partial x}\right) .$$

Direct comparison shows that these terms are neglegible compared with the other terms of Eq. (6.18), provided that $k^2 v_{os}^2 \ll \omega_{pe}^2$,

Lastly, we use the high frequency component of the continuity equation to simplify the thermal correction term in Eq. (6.18). In particular,

$$\frac{\partial n_e^h}{\partial t} + n_0 \frac{\partial u_e^h}{\partial x} + v_{os}\frac{\partial n_e^\ell}{\partial x} = 0 ,$$

which gives

$$\frac{3v_e^2}{n_0}\frac{\partial^2 n_e^h}{\partial t \partial x} = -3v_e^2\frac{\partial^2 u_e^h}{\partial x^2} - \frac{3v_e^2}{n_0} v_{os}\frac{\partial^2 n_e^\ell}{\partial x^2} .$$

The second term on the right hand side can be neglected compared to $(4\pi e^2/m)n_e^\ell v_{os}$, provided $k^2 \lambda_{De}^2 \ll 1$. Hence we finally obtain

$$\left(\frac{\partial^2}{\partial t^2} + \nu_e \frac{\partial}{\partial t} + \omega_{pe}^2 - 3v_e^2\frac{\partial^2}{\partial x^2}\right) u_e^h = -\frac{4\pi e^2}{m} n_e^\ell v_{os} . \quad (6.19)$$

To obtain an equation for the low frequency fluctuations, we must consider both the electron and ion responses. If we neglect electron inertia

$(Z\,m \ll M)$ and use an isothermal equation of state, the low frequency component of the force equation becomes

$$\frac{e}{m}\,E^\ell = -\frac{\partial}{\partial x}(v_{os}u_e^h) - \frac{v_e^2}{n_0}\frac{\partial n_e^\ell}{\partial x}\,. \tag{6.20}$$

In other words, the low frequency electric field transmits the ponderomotive force and the electron pressure to the ions.

The fluid equations for the ions with mass M and charge Ze are

$$\frac{\partial n_i}{\partial t} + \frac{\partial}{\partial x}(n_i u_i) = 0 \tag{6.21}$$

$$\frac{\partial u_i}{\partial t} + u_i\frac{\partial u_i}{\partial x} = \frac{Z\,e}{M}E - \frac{1}{M\,n_{0i}}\frac{\partial p_i}{\partial x} - \nu_i u_i \tag{6.22}$$

$$\frac{p_i}{n_i^3} = \text{constant}\,, \tag{6.23}$$

where we have included collisions with frequency ν_i to model either collisional or Landau damping of the ion fluctuations. Neglecting the response of the massive ions to the high frequency fields, we linearize these equations by assuming

$$\begin{aligned} n_i &= n_{0i} + n_i^\ell \\ u_i &= u_i^\ell\,. \end{aligned} \tag{6.24}$$

The linearized continuity and force equations become

$$\frac{\partial n_i^\ell}{\partial t} + n_{0i}\frac{\partial u_i^\ell}{\partial x} = 0 \tag{6.25}$$

$$\frac{\partial u_i^\ell}{\partial t} = \frac{Z\,e}{M}E^\ell - \frac{3v_i^2}{n_{0i}}\frac{\partial n_i^\ell}{\partial x} - \nu_i\,u_i^\ell\,. \tag{6.26}$$

We next take $\partial/\partial t$ of Eq. (6.25), $\partial/\partial x$ of Eq. (6.26), and eliminate the common term $\partial^2 u_i^\ell/\partial x\partial t$ to obtain

$$\frac{\partial^2 n_i^\ell}{\partial t^2} + \nu_i\frac{\partial n_i^\ell}{\partial t} + \frac{Z\,e\,n_{0i}}{M}\frac{\partial E^\ell}{\partial x} - 3v_i^2\frac{\partial^2 n_i^\ell}{\partial x^2} = 0\,. \tag{6.27}$$

Substituting for E from Eq. (6.20) and noting that $Z\,n_i^\ell \simeq n_e^\ell$ then gives

$$\frac{\partial^2 n_e^\ell}{\partial t^2} + \nu_i\frac{\partial n_e^\ell}{\partial t} - v_s^2\frac{\partial^2 n_e^\ell}{\partial x^2} = \omega_{\text{pi}}^2\,\frac{m}{4\pi e^2}\,v_{os}\frac{\partial^2 u_e^h}{\partial x^2}\,, \tag{6.28}$$

where $v_s^2 = (Z\theta_e + 3\theta_i)/M$ and $\omega_{\text{pi}}^2 = 4\pi n_0 e^2 Z/M$.

6.5   DISPERSION RELATION

The coupled equations for u_e^h and n_e^ℓ describe the feedback which leads to instability. To derive the dispersion relation, we assume that u_e^h and n_e^ℓ vary as $\exp(ikx - i\omega t)$ and represent $v_{os}(t)$ as

$$v_{os}(t) = v_{os} \left[\frac{\exp(i\omega_0 t) + \exp(-i\omega_0 t)}{2} \right],$$

where $v_{os} = eE_0/m\omega_0$. The Fourier transforms in space and time of Eqs. (6.19) and (6.28) give

$$\left(\omega^2 + i\omega\nu_e - \omega_{ek}^2 \right) u_{ek}^h(\omega) = \qquad\qquad\qquad (6.29)$$

$$\frac{4\pi e^2}{m} \frac{v_{os}}{2} \left[n_{ek}^\ell(\omega - \omega_0) + n_{ek}^\ell(\omega + \omega_0) \right]$$

$$\left(\omega^2 + i\omega\nu_i - k^2 v_s^2 \right) n_{ek}^\ell(\omega) = \qquad\qquad\qquad (6.30)$$

$$\frac{\omega_{pi}^2 m}{4\pi e^2} \frac{k^2 v_{os}}{2} \left[u_{ek}^h(\omega - \omega_0) + u_{ek}^h(\omega + \omega_0) \right],$$

where $\omega_{ek}^2 = \omega_{pe}^2 + 3k^2 v_e^2$. Choosing ω as low frequency, we use Eq. (6.29) to eliminate $u_{ek}^h(\omega \pm \omega_0)$ from Eq. (6.30). Noting that $|\omega \pm 2\omega_0| \simeq 2\omega_0$, we neglect as very off-resonant the terms $n_{ek}^\ell(\omega + 2\omega_0)$ and $n_{ek}^\ell(\omega - 2\omega_0)$. Hence we obtain the dispersion relation

$$\omega^2 + i\omega\nu_i - k^2 v_s^2 = \frac{\omega_{pi}^2 k^2 v_{os}^2}{4} \left[\frac{1}{\bar\epsilon(k, \omega - \omega_0)} + \frac{1}{\bar\epsilon(k, \omega + \omega_0)} \right], \quad (6.31)$$

where $\bar\epsilon(k, \omega) = \omega^2 + i\nu_e\omega - \omega_{ek}^2$.

This dispersion relation can be simplified considerably. First, note that

$$\bar\epsilon(k, \omega \pm \omega_0) = \left(\omega \pm \omega_0 + \omega_{ek} \right)\left(\omega \pm \omega_0 - \omega_{ek} \right) + i\left(\omega \pm \omega_0 \right)\nu_e.$$

Defining $\delta = \omega_0 - \omega_{ek}$, approximating $\omega_0 + \omega_{ek} \simeq 2\omega_0$, and assuming that $\omega \ll \omega_0$ gives

$$\bar\epsilon(k, \omega \pm \omega_0) = \pm 2\omega_0 (\omega \pm \delta) \pm i\nu_e\omega_0. \qquad (6.32)$$

Substituting Eq. (6.32) into (6.31) and rearranging, we obtain

$$\left(\omega^2 + i\omega\nu_i - k^2 v_s^2\right)\left[\left(\omega + \frac{i\nu_e}{2}\right)^2 - \delta^2\right] + \omega_{pi}^2 \frac{k^2 v_{os}^2}{4}\frac{\delta}{\omega_0} = 0 . \quad (6.33)$$

Both the oscillating two stream and ion-acoustic decay instability are readily determined from this dispersion relation. Let us first look for a purely growing instability, which corresponds to the oscillating two stream. For $\omega = i\gamma$, Eq.(6.33) becomes

$$\left(\gamma^2 + k^2 v_s^2\right)\left[\left(\gamma + \frac{\nu_e}{2}\right)^2 + \delta^2\right] + \frac{\omega_{pi}^2 k^2 v_{os}^2}{4}\frac{\delta}{\omega_0} = 0 , \quad (6.34)$$

where ν_i has been taken to be zero as is appropriate for ion Landau damping of a wave with zero phase velocity. Clearly $\gamma > 0$ requires that $\delta < 0$ i.e., $\omega_0 < \omega_{ek}$ as expected from the physical picture of this instability which we discussed in Section 6.3.

Expressions for the maximum growth rate are readily obtained in both the weak and strong growth limits. In the weak growth limit, $\gamma \ll kv_s$ and Eq. (6.34) becomes

$$\left(\gamma + \frac{\nu_e}{2}\right)^2 + \delta^2 + \frac{1}{4}\left(\frac{v_{os}}{v_e}\right)^2 \omega_0 \delta = 0 . \quad (6.35)$$

We find the mismatch δ (and hence wavenumber k) which corresponds to maximum growth by taking the derivative $\partial/\partial\delta$ of Eq. (6.35) and setting $\partial\gamma/\partial\delta = 0$. Hence $\delta = -(v_{os}/v_e)^2 \omega_0/8$. Substituting δ into Eq. (6.35) then gives the maximum growth rate, which is

$$\gamma = \frac{1}{8}\left(\frac{v_{os}}{v_e}\right)^2 \omega_0 - \frac{\nu_e}{2} . \quad (6.36)$$

Due to the damping of the plasma wave, the amplitude of the pump field must exceed a certain threshold value for net growth to occur. This threshold value is simply given by the condition $\gamma = 0$:

$$\left(\frac{v_{os}}{v_e}\right)^2_{TH} = 4\frac{\nu_e}{\omega_0} . \quad (6.37)$$

Let us next consider the regime of very strong growth. For growth rate $\gamma \gg (kv_s, \nu_e)$, Eq. (6.34) becomes

$$\gamma^2\left(\gamma^2 + \delta^2\right) + \frac{\omega_{pi}^2 k^2 v_{os}^2}{4}\frac{\delta}{\omega_0} = 0 . \quad (6.38)$$

We again take $\partial/\partial\delta$ of Eq. (6.38) and set $\partial\gamma/\partial\delta = 0$ to estimate the δ which corresponds to the maximum growth rate. After straightforward algebra, we obtain

$$\gamma \simeq \left(\frac{Zm}{M}\frac{k^2 v_{os}^2}{8}\omega_0\right)^{1/3} , \qquad (6.39)$$

for a mismatch $\delta = -\gamma$.

The mentioned ion-acoustic-decay instability is also readily obtained from Eq. (6.33). We first examine the weak growth regime, assuming that $\gamma \ll kv_s$. Maximum growth clearly occurs when both the ion acoustic wave and the electron plasma wave are nearly resonant. Hence we take $\omega = kv_s + i\gamma$ and choose $\delta = \omega_0 - \omega_{ek} = kv_s$. Substitution into Eq. (6.33) then gives a quadratic equation for the growth rate γ:

$$4\gamma^2 + 2\gamma(\nu_i + \nu_e) + \nu_i\nu_e - \frac{1}{4}\left(\frac{v_{os}}{v_e}\right)^2 k v_s\omega_0 = 0 , \qquad (6.40)$$

where we have used $\gamma \ll kv_s$. Growth again requires that the pump field exceed a threshold value, which is obtained by the condition that $\gamma = 0$:

$$\left(\frac{v_{os}}{v_e}\right)_{TH}^2 = 4\frac{\nu_e}{\omega_0}\frac{\nu_i}{kv_s} . \qquad (6.41)$$

For a growth rate much greater than either collision frequency but still much less than the ion acoustic frequency,

$$\gamma = \frac{1}{4}\frac{v_{os}}{v_e}\sqrt{k v_s\omega_0} . \qquad (6.42)$$

For large amplitude pump fields, the frequency of the ion wave can be determined by the pump field intensity. In this limit, the ion wave is called a quasi-mode since it is not a mode of the undriven plasma. Assuming that $|\omega| \gg kv_s$ and ignoring the damping terms, we return to Eq.(6.33) to obtain

$$\omega^2 (\omega^2 - \delta^2) + \omega_{pi}^2\frac{k^2 v_{os}^2}{4}\frac{\delta}{\omega} = 0 . \qquad (6.43)$$

To solve Eq. (6.43), we take $\omega = |\omega|\exp(i\phi)$ and and $\delta = \alpha|\omega|$, where α is a parameter to be varied to maximize the growth. The imaginary part of Eq. (6.43) then gives $\sin\phi = \frac{1}{2}\sqrt{2 - \alpha^2}$. The real part of this equation gives

$$|\omega| = \left(\frac{\omega_{pi}^2 k^2 v_{os}^2 \alpha}{4\omega_0}\right)^{1/3}$$

The growth rate, $\gamma = |\omega| \sin \phi$, maximizes when $\alpha = 1/\sqrt{2}$. Hence we obtain the real part of the frequency (ω_r) and the growth rate:

$$\omega_r \simeq \frac{\sqrt{5}}{2} \left(\frac{\omega_{\mathrm{pi}}^2}{\omega_0} \frac{k^2 v_{\mathrm{os}}^2}{16} \right)^{1/3}$$

$$\gamma \simeq \frac{\sqrt{3}}{2} \left(\frac{\omega_{\mathrm{pi}}^2}{\omega_0} \frac{k^2 v_{\mathrm{os}}^2}{16} \right)^{1/3} . \tag{6.44}$$

6.6 INSTABILITY THRESHOLD DUE TO SPATIAL INHOMOGENEITY

The threshold for instability is often determined not by collisions but by spatial inhomogeneties. In a plasma with a gradient in density, the oscillating-two-steam and ion acoustic decay instabilities are excited only over a region of limited size i.e., where $\omega_0 \sim \omega_{\mathrm{pe}}$. There is then a loss mechanism, since the unstable waves can propagate out of the region in which they are excited. Let us conclude our discussion of these instabilities by estimating the effect of a density gradient on the threshold [10].

We start by considering a plasma wave driven unstable by the electric vector of a light wave which is normally incident onto an inhomogeneous plasma. For simplicity, we consider only the oscillating two stream instability and assume that the plasma density varies linearly near the critical density with scale length L. Where the excitation is strongest, the plasma wave has a wavenumber $k_\parallel$ aligned with the electric vector of the light wave. However, at a lower density, the wave vector must develop a component (k_z) down the density gradient so that the increase in the frequency due to the thermal correction balances the decrease due to lower density. Hence

$$3\, k_z^2\, v_e^2 = \omega_{\mathrm{pe}}^2\, \frac{z}{L} , \tag{6.45}$$

where $z = 0$ corresponds to the place where $k = k_\parallel$. As k_z increases, the efficiency of the coupling between the light wave and the plasma wave decreases, since the plasma wave begins to propagate more and more in a direction orthogonal to the electric vector of the pump field. If we estimate the size (ℓ_{INT}) of the interaction region by the condition $k_z \simeq k_\parallel$,

$\ell_{INT} \simeq (3v_e^2 k_\parallel^2 L/\omega_{pe}^2)$. The time that it takes for wave energy to propagate out of this region is

$$\tau = \int_0^{\ell_{INT}} \frac{dz}{v_{gz}} , \qquad (6.46)$$

where v_{gz} is the component of the group velocity of the plasma wave down the density gradient. Noting that $v_{gz} = 3k_z v_e^2/\omega_{pe}$ and using Eq. (6.45) for k_z, we readily obtain $\tau = 2k_\parallel L/\omega_{pe}$. The effective damping rate ($\nu = 1/\tau$) becomes $\nu/\omega_{pe} = 1/2k_\parallel L$. Substituting this damping into Eq. (6.37) for the threshold gives

$$\left(\frac{v_{os}}{v_e} \right)_{TH}^2 \simeq \frac{2}{k_\parallel L} . \qquad (6.47)$$

6.7   EFFECT OF INCOHERENCE IN THE PUMP WAVE

Finally let's note that either temporal or spatial incoherence in the pump wave will reduce the instability growth rate. As a simple example, consider the ion-acoustic decay instability driven by a spatially homogeneous pump field with a frequency near ω_{pe} and with a random modulation in its amplitude [11]. In particular, we let

$$E = E_0 \, \alpha(t) \cos \omega t ,$$

where E is the electric field, ω is the frequency, and $\alpha(t)$ is a stochastic variable with a zero mean and a variance of unity. When the growth rate is much less than the ion acoustic frequency, the amplitude ($\tilde{f}$) of an unstable wave can be represented by terms of the form

$$\tilde{f} = \beta \exp\left[\gamma_0 \int_0^t \alpha(t') dt' \right] . \qquad (6.48)$$

Here γ_0 is the growth rate in the absence of amplitude modulation, damping has been neglected, and β is a constant determined by the initial conditions. If we assume that $\alpha(t)$ is Gaussian,

$$\langle \tilde{f} \rangle = \beta \exp\left[\frac{\gamma_0^2}{2} \int_0^t dt' \int_0^t dt'' \, \langle \alpha(t')\alpha(t'') \rangle \right] , \qquad (6.49)$$

where the brackets denote an average. The effective bandwidth $\Delta\omega$ is defined via the autocorrelation function:

$$\frac{1}{\Delta\omega} = \int_0^\infty d\tau \, \langle \alpha(t)\alpha(t+\tau) \rangle .$$

If $\Delta\omega \gg \gamma_0$, Eq. (6.49) gives

$$\langle \tilde{f} \rangle = \beta \exp\left(\frac{\gamma_0^2 t}{\Delta\omega} \right) . \tag{6.50}$$

Hence the growth rate is reduced by the ratio $\gamma_0/\Delta\omega$.

This reduction in the growth rate is readily understood. The intensity of the pump wave is distributed over a bandwidth $\Delta\omega$, and the resonance width of the instability is the growth rate. Hence when $\Delta\omega \gg \gamma_0$, only some fraction of the pump wave resonantly couples to any two given unstable waves. A random modulation in the phase of the pump waves leads to the same reduction in the growth rate [12,13]. In general, a spread in the wave vectors of the pump field or even random turbulence in the plasma [14] can also limit the coherence and contribute to the effective bandwidth.

References

1. Dawson, J. M. and C. Oberman, Effect of ion correlations on high-frequency plasma conductivity, *Phys. Fluids* **6**, 394 (1963).

2. Faehl, R. J. and W. L. Kruer, Laser light absorption by short wavelength ion turbulence, *Phys. Fluids* **20**, 55 (1977).

3. Chen, F. F. *Introduction to Plasma Physics*. Plenum Press, New York, 1974.

4. Silin, V. P., Parametric resonances in a plasma, *Sov. Phys. JETP* **21**, 1127 (1965).

5. DuBois, D. F. and M. V. Goldman, Radiation-induced instability of electron plasma waves, *Phys. Rev. Letters* **14**, 544 (1965).

6. Nishikawa, K., Parametric excitation of coupled waves I. General formulation, *J. Phys. Soc. Japan*, **24**, 1152 (1968).

7. Kaw, P. K. and J. M. Dawson, Laser-induced anomalous heating of a plasma, *Phys. Fluids* **12**, 2586 (1969).

8. Sanmartin, J. R., Electrostatic plasma instabilities excited by high-frequency electric field, *Phys. Fluids* **13**, 1533 (1970).

9. Mima, K. and K. Nishikawa, Parametric instabilities and wave dissipation in plasmas; in *Handbook of Plasma Physics*, Vol. 2, (A. A. Galeev and R. N. Sudan, eds.), p.451–517. North Holland Physics Publishing, Amsterdam, 1984.

10. Perkins, F. W. and J. Flick, Parametric instabilities in inhomogeneous plasmas, *Phys. Fluids* **14**, 2012 (1971).

11. Thomson, J. J., W. L. Kruer, S. E. Bodner and J. S. DeGroot, Parametric instability thresholds and their control, *Phys. Fluids* **17**, 849 (1974).

12. Valeo, E. J. and C. Oberman, Model of parametric excitation by an imperfect pump, *Phys. Rev. Lett.* **30**, 1035 (1973).

13. Thomson, J. J. and J. I. Karush, Effect of finite-bandwidth driver on the parametric instability, *Phys. Fluids* **17**, 1608 (1974).

14. Williams, E. A., J. R. Albritton, and M. N. Rosenbluth, Effect of spatial turbulence on parametric instabilities, *Phys. Fluids* **22**, 139 (1979).

CHAPTER 7

Stimulated
Raman
Scattering

An important class of instabilities involves the coupling of a large amplitude light wave into a scattered light wave plus either an electron plasma wave (the Raman instability) or an ion acoustic wave (the Brillouin instability). In this Chapter, we will consider the Raman instability and a related instability in which a light wave couples into two electron plasma waves. In the next chapter, we will discuss the Brillouin instability and a related instability which can lead a beam of light to break up into filaments.

The Raman instability can be most simply characterized as the resonant decay of an incident photon into a scattered photon plus an electron plasma wave (or plasmon). The frequency and wave number matching conditions then are

$$\omega_0 = \omega_s + \omega_{ek}$$
$$\mathbf{k}_0 = \mathbf{k}_s + \mathbf{k},$$

$$(7.1)$$

where ω_0 (ω_s) and $\mathbf{k}_0$ $(\mathbf{k}_s)$ are the frequency and wave number of the incident (scattered) light wave, and ω_{ek} $(\mathbf{k})$ is the frequency (wavenumber) of the electron plasma wave. Since the minimum frequency of a light wave in a plasma is ω_{pe}, the electron plasma frequency, it is clear that this instability requires that $\omega_0 \gtrsim 2\omega_{\mathrm{pe}}$ i.e., $n \lesssim n_{\mathrm{cr}}/4$, where n is the plasma density and n_{cr} is the critical density.

In this process, part of the incident energy is scattered, and part is deposited into the electron plasma wave. If we simply multiply the frequency matching condition by $\hbar$ (Planck's constant) and note that $\hbar\omega$ is the energy of a photon or plasmon, it is clear that for each photon undergoing this process, the fraction of its energy transferred to the plasma wave is (ω_{ek}/ω_0). This portion of the energy will heat the plasma as the electron plasma wave damps. As we will see, this electron plasma wave can have a very high phase velocity (of order the velocity of light) and so can produce very energetic electrons when it damps. Since such electrons can preheat the fuel in laser fusion applications, the Raman instability is a particularly significant concern.

The physics of the Raman instability is straight forward. Consider a light wave with electric field amplitude $\mathbf{E}_L$ propagating through a plasma whose density is rippled along the direction of propagation by the density fluctuation δn associated with an electron plasma wave. Since the electrons are oscillating in the light wave with the velocity $\mathbf{v}_L = e\mathbf{E}_L/m\omega_0$, a transverse current $\delta\mathbf{J} = -e\,\mathbf{v}_L\,\delta n$ is generated. If the wave numbers and frequencies are properly matched, this transverse current generates a scattered light wave with an amplitude $\delta\mathbf{E}$. In turn, this scattered light wave interferes with the incident light to produce a variation in the wave pressure: $\nabla(E^2/8\pi) = \nabla(\mathbf{E}_L \cdot \delta\mathbf{E})/4\pi$. Variations in wave pressure act just like variations in the ordinary kinetic pressure i.e., plasma is pushed from regions of high pressure to regions of low pressure and vice versa, and a density fluctuation is generated. Due to this feed-back loop, an instability is possible. A small density fluctuation leads to a transverse current which generates a small scattered light wave, which can in turn reinforce the density fluctuation via a variation in the wave pressure.

7.1 INSTABILITY ANALYSIS

The coupled equations describing the Raman instability can be readily derived [1–3]. For clarity, let us consider a light wave propagating through a plasma with a uniform density and temperature. It is convenient to express the electric and magnetic fields in terms of the vector potential $\mathbf{A}$ and the electrostatic potential ϕ, where $\mathbf{B} = \nabla \times \mathbf{A}$ and $\mathbf{E} = -c^{-1}\partial\mathbf{A}/\partial t - \nabla\phi$. We begin with Ampere's law

$$\nabla \times \mathbf{B} = \frac{4\pi}{c}\mathbf{J} + \frac{1}{c}\frac{\partial\mathbf{E}}{\partial t}. \tag{7.2}$$

Substituting for $\mathbf{B}$ and $\mathbf{E}$ and choosing $\nabla \cdot \mathbf{A} = 0$, we obtain

$$\left(\frac{1}{c^2}\frac{\partial^2}{\partial t^2} - \nabla^2\right)\mathbf{A} = \frac{4\pi}{c}\mathbf{J} - \frac{1}{c}\frac{\partial}{\partial t}\nabla\phi. \tag{7.3}$$

We next separate the current density $\mathbf{J}$ into a transverse part $\mathbf{J}_t$ (associated with the light waves) and a longitudinal part $\mathbf{J}_\ell$ (associated with the electrostatic plasma wave). The longitudinal part of $\mathbf{J}$ can be related to $\nabla\phi$ via Poisson's equation and the equation for conservation of charge:

$$\nabla^2\phi = -4\pi\rho \tag{7.4}$$

$$\frac{\partial\rho}{\partial t} + \nabla \cdot \mathbf{J} = 0, \tag{7.5}$$

where ρ is the charge density. In particular, taking $\partial/\partial t$ of Eq. (7.4) and substituting for $\partial\rho/\partial t$ from Eq. (7.5) gives

$$\nabla \cdot \left(\frac{\partial}{\partial t}\nabla\phi - 4\pi\mathbf{J}\right) = 0. \tag{7.6}$$

Since $\nabla \cdot \mathbf{J}_t = 0$, we then obtain

$$\frac{\partial}{\partial t}\nabla\phi = 4\pi\mathbf{J}_\ell. \tag{7.7}$$

Hence Eq. 7.3 becomes

$$\left(\frac{1}{c^2}\frac{\partial^2}{\partial t^2} - \nabla^2\right)\mathbf{A} = \frac{4\pi}{c}\mathbf{J}_t. \tag{7.8}$$

If we restrict ourselves to the condition $\mathbf{A} \cdot \nabla n_e = 0$, the transverse current can be simply expressed as $\mathbf{J}_t = -n_e e\mathbf{u}_t$. Here $\mathbf{u}_t$ is the oscillation velocity of an electron in the electric field of the light wave and n_e is the electron density. For $|\mathbf{u}_t| \ll c$, $\mathbf{u}_t = e\mathbf{A}/mc$ since

$$\frac{\partial\mathbf{u}_t}{\partial t} = -\frac{e}{m}\mathbf{E}_t = \frac{e}{mc}\frac{\partial\mathbf{A}}{\partial t}. \tag{7.9}$$

Hence, we obtain an equation for the propagation of a light wave in a plasma:

$$\left(\frac{\partial^2}{\partial t^2} - c^2 \nabla^2 \right) \mathbf{A} = - \frac{4\pi e^2}{m} n_e \mathbf{A} . \tag{7.10}$$

The scattering of a large amplitude light wave ($\mathbf{A}_L$) by a small amplitude density fluctuation ($\tilde{n}_e$) is easily determined by substituting into Eq. 7.10 for $\mathbf{A} = \mathbf{A}_L + \tilde{\mathbf{A}}$ and for $n = n_0 + \tilde{n}_e$ where n_0 is the uniform background plasma density. We then obtain

$$\left(\frac{\partial^2}{\partial t^2} - c^2 \nabla^2 + \omega_{\mathrm{pe}}^2 \right) \tilde{\mathbf{A}} = - \frac{4\pi e^2}{m} \tilde{n}_e \mathbf{A}_L . \tag{7.11}$$

The right hand side is simply the transverse current ($\propto \tilde{n}_e \mathbf{v}_L$) which produces the scattered light wave ($\tilde{\mathbf{A}}$).

To derive an equation for the density fluctuation associated with the electron plasma wave, we treat the massive ions as a fixed, neutralizing background and describe the electrons as a warm fluid. The continuity and force equations then are

$$\frac{\partial n_e}{\partial t} + \nabla \cdot \left(n_e \mathbf{u}_e \right) = 0 \tag{7.12}$$

$$\frac{\partial \mathbf{u}_e}{\partial t} + \mathbf{u}_e \cdot \nabla \mathbf{u}_e = \frac{-e}{m} \left(\mathbf{E} + \frac{\mathbf{u}_e \times \mathbf{B}}{c} \right) - \frac{\nabla p_e}{n_e m} , \tag{7.13}$$

where n_e, $\mathbf{u}_e$ and p_e are the density, velocity and pressure of the electron fluid. (As we have shown in Chapter 1, these equations are readily derived as the first two moments of the Vlasov equation.) Separating the velocity into longitudinal and transverse components ($\mathbf{u}_e = \mathbf{u}_L + e\mathbf{A}/mc$), substituting into Eq. (7.13), and using a standard vector identity gives

$$\frac{\partial \mathbf{u}_L}{\partial t} = \frac{e}{m} \nabla \phi - \frac{1}{2} \nabla \left(\mathbf{u}_L + \frac{e\mathbf{A}}{mc} \right)^2 - \frac{\nabla p_e}{n_e m} . \tag{7.14}$$

The second term on the right hand side is the ponderomotive force and is proportional to the gradient of the intensity of both the longitudinal and transverse components of the electric field.

We now use the adiabatic equation of state ($p_e/n_e^3 = $ constant) and linearize Eqs. (7.12) and (7.14.) In particular, we take $\mathbf{u}_L = \tilde{\mathbf{u}}$, $\quad n_e = $

$n_0 + \tilde{n}_e$, $\mathbf{A} = \mathbf{A}_L + \tilde{\mathbf{A}}$ and $\phi = \tilde{\phi}$ where the tilde denotes an infinitesimal quantity. Then

$$\frac{\partial \tilde{n}_e}{\partial t} + n_0 \nabla \cdot \tilde{\mathbf{u}} = 0 \tag{7.15}$$

$$\frac{\partial \tilde{\mathbf{u}}}{\partial t} = \frac{e}{m} \nabla \tilde{\phi} - \frac{e^2}{m^2 c^2} \nabla \left(\mathbf{A}_L \cdot \tilde{\mathbf{A}} \right) - \frac{3 v_e^2}{n_0} \nabla \tilde{n}_e , \tag{7.16}$$

where v_e is the electron thermal velocity. Taking a time derivative of Eq. (7.15), then a divergence of Eq. (7.16), and finally eliminating the term $\partial(\nabla \cdot \tilde{\mathbf{u}}_L)/\partial t$ gives

$$\left(\frac{\partial^2}{\partial t^2} + \omega_{\text{pe}}^2 - 3 v_e^2 \nabla^2 \right) \tilde{n}_e = \frac{n_0 e^2}{m^2 c^2} \nabla^2 \left(\mathbf{A}_L \cdot \tilde{\mathbf{A}} \right) . \tag{7.17}$$

Here we have also made use of Poisson's equation ($\nabla^2 \tilde{\phi} = 4\pi e \tilde{n}_e$) to eliminate $\tilde{\phi}$. This equation describes the generation of a fluctuation in the electron density by variations in the intensity of the electromagnetic waves.

7.2 DISPERSION RELATION

Equations (7.11) and (7.17) describe the coupling of the electrostatic and electromagnetic waves discussed in the introduction to this chapter. To derive the dispersion relation for the Raman instability, we here take $\mathbf{A}_L = \mathbf{A}_0 \cos(\mathbf{k}_0 \cdot \mathbf{x} - \omega_0 t)$ and Fourier-analyze these equations:

$$\left(\omega^2 - k^2 c^2 - \omega_{\text{pe}}^2 \right) \tilde{\mathbf{A}}(\mathbf{k}, \omega) = \tag{7.18}$$

$$\frac{4\pi e^2}{2m} \mathbf{A}_0 \left[\tilde{n}_e(k - k_0, \omega - \omega_0) + \tilde{n}_e(k + k_0, \omega + \omega_0) \right]$$

$$\left(\omega^2 - \omega_{ek}^2 \right) \tilde{n}_e(\mathbf{k}, \omega) = \tag{7.19}$$

$$\frac{k^2 e^2 n_0}{2 m^2 c^2} \mathbf{A}_0 \cdot \left[\tilde{\mathbf{A}}(k - k_0, \omega - \omega_0) + \tilde{\mathbf{A}}(k + k_0, \omega + \omega_0) \right] ,$$

where $\omega_{ek} = \left(\omega_{\text{pe}}^2 + 3 k^2 v_e^2 \right)^{1/2}$ is the Bohm-Gross frequency and ω_0 and k_0 are the frequency and wave number of the large amplitude light wave. We next use Eq. (7.18) to eliminate $\tilde{\mathbf{A}}$ from Eq. (7.19). Taking $\omega \simeq \omega_{\text{pe}}$

and neglecting the terms $\tilde{n}_e(k - 2k_0, \omega - 2\omega_0)$ and $\tilde{n}_e(k + 2k_0, \omega + 2\omega_0)$ as very nonresonant, we obtain the dispersion relation:

$$\omega^2 - \omega_{ek}^2 = \frac{\omega_{pe}^2 k^2 v_{os}^2}{4} \left[\frac{1}{D(\omega - \omega_0, \mathbf{k} - \mathbf{k}_0)} + \frac{1}{D(\omega + \omega_0, \mathbf{k} + \mathbf{k}_0)} \right] . \quad (7.20)$$

Here $D(\omega, k) = \omega^2 - k^2 c^2 - \omega_{pe}^2$ and v_{os} is the oscillatory velocity of an electron in the large amplitude light wave.

The instability growth rates are readily found from Eq. (7.20). For back or sidescatter, we can neglect the upshifted light wave as nonresonant, giving

$$(\omega^2 - \omega_{ek}^2) \left[(\omega - \omega_0)^2 - (\mathbf{k} - \mathbf{k}_0)^2 c^2 - \omega_{pe}^2 \right] = \frac{\omega_{pe}^2 k^2 v_{os}^2}{4} . \quad (7.21)$$

We take $\omega = \omega_{ek} + \delta\omega$, where $\delta\omega \ll \omega_{ek}$, and note that maximum growth occurs when the scattered light wave is also resonant i.e., when

$$\left(\omega_{ek} - \omega_0 \right)^2 - \left(\mathbf{k} - \mathbf{k}_0 \right)^2 c^2 - \omega_{pe}^2 = 0 . \quad (7.22)$$

Then, $\delta\omega = i\gamma$, where

$$\gamma = \frac{k \, v_{os}}{4} \left[\frac{\omega_{pe}^2}{\omega_{ek} (\omega_0 - \omega_{ek})} \right]^{1/2} . \quad (7.23)$$

The wave number k is determined by Eq. (7.22). For example, for direct backscatter where the growth rate maximizes,

$$k = k_0 + \frac{\omega_0}{c} \left(1 - \frac{2\omega_{pe}}{\omega_0} \right)^{1/2} . \quad (7.24)$$

The wave number starts from $k = 2k_0$ for $n \ll n_{cr}/4$, and goes to $k = k_0$ for $n \sim n_{cr}/4$, as is apparent from the matching condition.

The wave number and growth rate are less for any 90° sidescatter $(k \simeq \sqrt{2}\, k_0$ for $n \ll n_{cr}/4)$. For the more general case of sidescatter in which $\mathbf{A} \cdot \nabla n \neq 0$, the growth rate is further reduced since the electric vectors of the incident and scattered light waves are no longer aligned. For example, it is apparent from Eq. (7.17) that the growth rate will vanish when $\tilde{\mathbf{A}} \cdot \mathbf{A}_0 = 0$. Hence sidescatter occurs preferentially out of the plane of polarization, the case we have treated.

For forward scatter at very low density, $k \ll \omega_0/c$. Both upshifted and downshifted light waves can now be nearly resonant i.e.,

$$D(\omega \pm \omega_0, \mathbf{k} \pm \mathbf{k}_0) \simeq 2(\omega_{\text{pe}} \pm \omega_0)\delta\omega \,,$$

where we have chosen $k = \omega_{\text{pe}}/c$ and let $\omega = \omega_{\text{pe}} + \delta\omega$, where $\delta\omega \ll \omega_{\text{pe}}$. Substituting into Eq. (7.20), we readily find the maximum growth rate $(\delta\omega = i\,\gamma)$:

$$\gamma \simeq \frac{\omega_{\text{pe}}^2}{2\sqrt{2}\,\omega_0}\,\frac{v_{\text{os}}}{c}\,. \tag{7.25}$$

Lastly, let us note that there is also a kinetic instability which represents stimulated Compton scattering by the electrons [4,5]. Now the electrostatic fluctuation is no longer a resonant mode of the plasma but rather a beat mode which interacts with the electrons. This instability can be readily derived from Eq. (7.20) if we replace $\omega^2 - \omega_{\text{pe}}^2$ by $\omega^2\,\epsilon(k,\omega)$ where $\epsilon(k,\omega)$ is the fully kinetic dielectric function. For a Maxwellian velocity distribution, the growth rate peaks when $\omega_0 \simeq \omega_s + k\,v_e$. Not surprisingly, the maximum growth rate is much less than that for the Raman instability, unless the plasma wave is heavily damped. The two processes then merge.

7.3 INSTABILITY THRESHOLDS

Damping of the unstable waves introduces a threshold intensity for instability generation. The simplest way to include the effect of damping is to add terms $\nu_s\,(\partial\tilde{\mathbf{A}}/\partial t)$ and $\nu_e\,(\partial\tilde{n}/\partial t)$ to Eqs. (7.11) and (7.17), where $\nu_s\,(\nu_e)$ is the energy damping rate for the scattered light wave (the electron plasma wave). The dispersion relation remains the same as Eq. (7.20) with the substitutions

$$\omega^2 - \omega_{\text{pe}}^2 \implies \omega\,(\omega + i\,\nu_e) - \omega_{\text{pe}}^2 \,,$$
$$D(\omega, k) \implies \omega\,(\omega + i\,\nu_s) - k^2 c^2 - \omega_{\text{pe}}^2 \,.$$

The instability analysis proceeds as before. For example, for back or sidescatter, we again retain only the down-shifted light wave, take $\omega = \omega_{ek} + i\gamma$, and choose k according to Eq. (7.22) to obtain maximum growth. Then we obtain

$$(\gamma + \gamma_e)\,(\gamma + \gamma_s) = \gamma_0^2 \,,$$

where γ_e and γ_s are the amplitude damping rates (half of the energy damping rates) and γ_0 is the growth rate in the absence of damping. The threshold condition due to damping then is

$$\gamma_0 \geq \sqrt{\gamma_e \gamma_s} \, . \tag{7.27}$$

As an example, we consider backscatter for $\omega_{pe}/\omega_0 \ll 1/2$ and assume only collisional damping. Substituting Eq. (7.23) into Eq. (7.27) then gives

$$\left(\frac{v_{os}}{c}\right)^2 > \left(\frac{\omega_{pe}}{\omega_0}\right)^2 \frac{\nu_{ei}^2}{\omega_0 \, \omega_{pe}} \, , \tag{7.28}$$

where ν_{ei} is the collision frequency discussed in Chapter 5. This threshold intensity can be quite low. In general, Landau damping of the plasma wave needs to be included, as will be discussed in Chapter 9.

In practice, the threshold intensity is usually determined by gradients in the plasma density rather than by damping. Let us conclude our discussion of the linear theory of the Raman instability with a heuristic calculation of the threshold in a plasma with a linear density profile. Plasma inhomogeneity limits the region over which three waves can resonantly interact, and propagation of wave energy out of this region introduces an effective dissipation which must be overcome. Noting that the wave numbers are now a function of position, let us define $\kappa = k_1(z) - k_2(z) - k_3(z)$. At some point $\kappa = 0$ (i.e., the waves are resonantly coupled), but away from this point a mismatch develops. The resonant coupling is spoiled when a significant phase shift develops. Hence we can estimate the size ℓ_{INT} of the interaction region by the condition $\int_0^{\ell_{INT}} \kappa \, dz \sim 1/2$. Taylor expanding about the matching point $(\kappa = \kappa(0) + \kappa' z)$ then gives $\ell_{INT} \sim 1/\sqrt{\kappa'}$. Propagation of wave energy out of this interaction region introduces an effective damping rate of approximately v_{gi}/ℓ_{INT}, where v_{gi} is the component of the group velocity of the ith wave along the gradient. Inserting these damping rates into Eq. (7.27) then gives the Rosenbluth criterion for $\exp(2\pi)$ amplification in a plasma with linear variation in κ:

$$\frac{\gamma^2}{|\kappa' \, v_{g1} \, v_{g2}|} \gtrsim 1 \, , \tag{7.29}$$

where 1 and 2 refer to the growing waves.

As an example, we consider Raman backscatter at $n \ll n_{cr}/4$. Since the wave number of the electron plasma wave depends more sensitively on density than do the wave numbers of the transverse waves, $\kappa' \simeq -\,\partial k/\partial x$ and $|v_{gp}\kappa'| \simeq \partial\omega_{\mathrm{pe}}/\partial x$. Neglecting temperature gradients and assuming a locally linear variation in density with a scale length $L = n/(\partial n/\partial x)$, $\partial\omega_{\mathrm{pe}}/\partial x \simeq \omega_{\mathrm{pe}}/2L$. Noting that $v_{g2} \simeq c$ and substituting into Eq. (7.29), we obtain the threshold condition:

$$\left(\frac{v_{\mathrm{os}}}{c}\right)^2 > \frac{2}{k_0\,L}\,. \tag{7.30}$$

In general, a more detailed treatment of the instability generation in inhomogeneous plasma is required. As the region of $n_{\mathrm{cr}}/4$ is approached, the group velocity of the scattered light wave decreases towards zero, and the WKB approximation fails. There the threshold becomes lower by a factor of $\sim (k_0\,L)^{1/3}$, which is roughly the maximum factor by which the group velocity of a light wave decreases in an inhomogeneous plasma. The threshold for Raman sidescatter is also lower than that given in Eq. (7.30) by a similar factor, since the sidescattered light wave is more weakly affected by the gradient in density. The threshold intensity is also substantially reduced at density maxima (where $\kappa' = 0$). In all these cases the instability can become absolute. The unstable waves do not then limit by convection but grow in time until nonlinear effects onset. An extensive discussion of the thresholds due to plasma inhomogeneity and the convective or absolute nature of the instability is given in the literature [6–11].

7.4   THE $2\,\omega_{\mathrm{pe}}$ INSTABILITY

Finally, let us briefly consider a related instability in which the laser light decays into two electron plasma waves [12–17]. The frequency and wavenumber matching conditions for this so-called $2\omega_{\mathrm{pe}}$ instability are

$$\begin{aligned} \omega_0 &= \omega_{ek1} + \omega_{ek2} \\ \mathbf{k}_0 &= \mathbf{k}_1 + \mathbf{k}_2\,, \end{aligned} \tag{7.31}$$

where ω_0 ($\mathbf{k}_0$) is the laser light frequency (wave number) and ω_{ek1} ($\mathbf{k}_1$) and ω_{ek2} ($\mathbf{k}_2$) are the frequencies (wave numbers) of the electron plasma waves. Since ω_{ek1} and ω_{ek2} are approximately ω_{pe}, this instability clearly takes place at a density $n \simeq n_{\mathrm{cr}}/4$. The $2\omega_{\mathrm{pe}}$ instability is a preheat concern, since electron plasma waves are generated.

To derive this instability, we can simply treat the ions as a fixed, neutralizing background and describe the electrons as a warm fluid. If we again express $\mathbf{u}_e = \mathbf{u}_L + \mathbf{v}_{os}$ where $\mathbf{v}_{os} = e\mathbf{A}_0/mc$ and linearize Eqs. (7.12) and (7.14), we obtain

$$\frac{\partial \tilde{n}_e}{\partial t} + n_0 \nabla \cdot \tilde{\mathbf{u}}_L + \mathbf{v}_{os} \cdot \nabla \tilde{n}_e = 0 \tag{7.32}$$

$$\frac{\partial \tilde{\mathbf{u}}_L}{\partial t} = \frac{e}{m} \nabla \tilde{\phi} - \frac{3 v_e^2}{n_0} \nabla \tilde{n}_e - \nabla(\mathbf{v}_{os} \cdot \tilde{\mathbf{u}}_L) , \tag{7.33}$$

where $\tilde{\mathbf{u}}_L$, $\tilde{n}_e$ and $\tilde{\phi}$ are treated as infinitesimal quantities. We next take a time derivative of Eq. (7.32), a divergence of Eq. (7.33), use Poisson's equation, and combine to eliminate the term $\partial(\nabla \cdot \tilde{\mathbf{u}}_L)/\partial t$. This gives

$$\frac{\partial^2 \tilde{n}_e}{\partial t^2} + (\omega_{pe}^2 - 3 v_e^2 \nabla^2)\tilde{n}_e + \frac{\partial(\mathbf{v}_{os} \cdot \nabla \tilde{n}_e)}{\partial t} - n_0 \nabla^2(\mathbf{v}_{os} \cdot \tilde{\mathbf{u}}_L) = 0 . \tag{7.34}$$

Representing $v_{os} = v_{os}[\exp(i\mathbf{k}_0 \cdot \mathbf{x} - i\omega_0 t) + \exp(-i\mathbf{k}_0 \cdot \mathbf{x} + i\omega_0 t)]/2$ and Fourier-analyzing Eq. (7.34) gives

$$(-\omega^2 + \omega_{ek}^2) \tilde{n}_e(k,\omega)$$
$$+ \frac{\omega}{2} \mathbf{k} \cdot \mathbf{v}_{os} \left[\tilde{n}_e(k - k_0, \omega - \omega_0) + \tilde{n}_e(k + k_0, \omega + \omega_0) \right] \tag{7.35}$$
$$+ \frac{n_0 k^2}{2} \mathbf{v}_{os} \cdot \left[\tilde{\mathbf{u}}_L(k - k_0, \omega - \omega_0) + \tilde{\mathbf{u}}_L(k + k_0, \omega + \omega_0) \right] = 0 .$$

An equation for $\tilde{n}_e(k - k_0, \omega - \omega_0)$ can be directly obtained from Eq. (7.35) by simply replacing k, ω with $k - k_0, \omega - \omega_0$. If we choose $\omega \sim \omega_{pe}$ and neglect as off-resonant any responses at $\omega + \omega_0$ or $\omega - 2\omega_0$, we easily obtain coupled equations for $\tilde{n}_e(k,\omega)$ and $\tilde{n}_e(k - k_0, \omega - \omega_0)$:

$$(-\omega^2 + \omega_{ek}^2) \tilde{n}_e(k,\omega)$$
$$+ \frac{\mathbf{v}_{os}}{2} \cdot \left[\omega \mathbf{k} \, \tilde{n}_e(k - k_0, \omega - \omega_0) \right. \tag{7.36}$$
$$\left. + n_0 k^2 \tilde{\mathbf{u}}_L(k - k_0, \omega - \omega_0) \right] = 0$$

$$\left[-(\omega - \omega_0)^2 + \omega_{ek-k_0}^2 \right] \tilde{n}_e(k - k_0, \omega - \omega_0) \tag{7.37}$$
$$+ \frac{\mathbf{v}_{os}}{2} \cdot \left[(\omega - \omega_0) \mathbf{k} \, \tilde{n}_e(k,\omega) + n_0(\mathbf{k} - \mathbf{k}_0)^2 \, \tilde{\mathbf{u}}_L(k,\omega) \right] = 0 .$$

Here we have noted that $\mathbf{k_0 \cdot v_{os}} = 0$. These equations describe the coupling of electron plasma waves with wave numbers k and $k - k_0$ by the laser light.

The Fourier-analyzed continuity equation is next used to approximate $\tilde{\mathbf{u}}_L$ in the coupling terms as

$$\tilde{\mathbf{u}}_L(k,\omega) \simeq \frac{\mathbf{k}}{k^2}\,\omega\,\frac{\tilde{n}_e(k,\omega)}{n_0}\,, \tag{7.38}$$

where we are neglecting the additional term involving v_{os} which would simply give a correction of order v_{os}^2. We then substitute Eq. (7.37) into Eq. (7.36) to obtain the dispersion relation

$$(\omega^2 - \omega_{ek}^2)\left[(\omega - \omega_0)^2 - \omega_{ek-k_0}^2\right] = \left[\frac{\mathbf{k \cdot v_{os}}\,\omega_{\text{pe}}[(\mathbf{k-k_0})^2 - k^2]}{2\,k\,|\mathbf{k-k_0}|}\right]^2 . \tag{7.39}$$

The coupling term has been simplified by approximating $\omega \simeq \omega_{\text{pe}}$ and $\omega - \omega_0 \simeq -\omega_{\text{pe}}$.

The growth rate is readily found by substituting $\omega = \omega_{ek} + i\gamma$ and invoking frequency matching. Then

$$\gamma \simeq \frac{\mathbf{k \cdot v_{os}}}{4}\left|\frac{(\mathbf{k-k_0})^2 - k^2}{k\,|\mathbf{k-k_0}|}\right| . \tag{7.40}$$

For $k \gg k_0$, the growth rate maximizes at $\gamma \simeq k_0 v_{os}/4$ for plasma waves propagating at $45°$ to both $\mathbf{k_0}$ and $\mathbf{v_{os}}$.

Either dissipation or plasma inhomogeneity introduce a threshold intensity for the instability. The collisional threshold is simply given by the condition that $\gamma = \nu_e/2$, where ν_e is the energy damping rate due to either electron-ion collisions or Landau damping. The threshold due to inhomogeneity is

$$\left(\frac{v_{os}}{v_e}\right)^2 \simeq \frac{12}{k_0 L}\,,$$

where L is the density scale length at $n_{\text{cr}}/4$. The inhomogeneous threshold for the $2\omega_{\text{pe}}$ instability is lower than that for the Raman instability at $n_{\text{cr}}/4$ unless the plasma is quite hot.

References

1. Drake, J. F., P. K. Kaw, Y. C. Lee, G. Schmidt, C. S. Liu, and M. N. Rosenbluth, Parametric instabilities of electromagnetic waves in plasmas, *Phys. Fluids* **17**, 778 (1974).

2. Forslund,D. W., J. M. Kindel and E. L. Lindman, Theory of stimulated scattering processes in laser-irradiated plasmas, *Phys. Fluids* **18**, 1002 (1975).

3. Thomson, J. J., Stimulated Raman scatter in laser fusion target chambers, *Phys. Fluids* **21**, 2082 (1978).

4. Ott, E., W. M. Manheimer and H. H. Klein, Stimulated Compton scattering and self-focusing in the outer regions of a laser fusion plasma, *Phys. Fluids* **17**, 1757 (1974).

5. Lin, A. T. and J. M. Dawson, Stimulated Compton scattering of electromagnetic waves in plasmas, *Phys. Fluids* **18**, 201 (1975).

6. Rosenbluth, M. N., Parametric instabilities in inhomogeneous media, *Phys. Rev. Letters* **29**, 565 (1972).

7. Nishikawa, K. and C. S. Liu, General formalism of parametric excitation; in *Advances in Plasma Physics*, Vol. 16, (A. Simon and W. Thompson, eds.), p.3–81. Wiley, New York, 1976.

8. Liu, C. S., Parametric instabilities in inhomogeneous unmagnetized plasmas, *ibid*, p.121–177.

9. Mima, K. and K. Nishikawa, Parametric instabilities and wave dissipation in plasmas; in *Handbook of Plasma Physics*, Vol. 2, (A. A. Galeev and R. N. Sudan, eds.), p.451–517. North-Holland, Amsterdam, 1984.

10. Koch, P. and E. A. Williams, Absolute growth of coupled forward and backward Raman scattering in inhomogeneous plasma, *Phys. Fluids* **27**, 2346 (1984).

11. Afeyan, B. B. and E. A. Williams, Stimulated Raman sidescattering with the effects of oblique incidence, *Phys. Fluids* **28**, 3397 (1985).

12. Goldman, M. V., Parametric plasmon-photon interactions, *Annals of Physics* **38**, 117 (1966).

13. Jackson, E. A., Parametric effects of radiation on a plasma, *Phys. Rev.* **153**, 235 (1967).

14. Liu, C. S. and M. N. Rosenbluth, Parametric decay of electromagnetic waves into two plasmons and its consequences, *Phys. Fluids* **19**, 967 (1976).

15. Simon, A., R. W. Short, E. A. Williams and T. Dewandre, On the inhomogeneous two-plasmon instability, *Phys. Fluids* **26**, 3107 (1983).

16. Lasinski, B. F. and A. B. Langdon, Linear theory of the $2\omega_{pe}$ instability in inhomogeneous plasmas, in *Lawrence Livermore Laboratory, UCRL–50021–77*, p.4–49 (1978).

17. Powers, L. V. and R. L. Berger, Kinetic theory of two plasmon decay, *Phys. Fluids* **27**, 242 (1984).

CHAPTER 8

Stimulated Brillouin Scattering

In this chapter we will consider the Brillouin instability, which involves the coupling of a large amplitude light wave into a scattered light wave plus an ion acoustic wave. The physics of this instability is analogous to that of the Raman instability, except that now the density fluctuation which provides the coupling to the scattered light wave is the density fluctuation associated with a low frequency ion acoustic wave. Our analysis will be sufficiently general to show another instability, which is also associated with the variations of plasma density induced by variations of light wave pressure. This latter instability is called the filamentation instability, since it can lead to the break-up of a light wave into filaments.

The Brillouin instability can be most simply characterized as the resonant decay of an incident photon with frequency ω_0 and wavenumber k_0 into a scattered photon with frequency ω_s and wavenumber k_s plus an ion acoustic phonon. The frequency and wave number matching conditions then are

$$\omega_0 = \omega_s + \omega$$
$$\mathbf{k}_0 = \mathbf{k}_s + \mathbf{k},$$

where now ω and $\mathbf{k}$ are the frequency and wave number of the ion acoustic wave. Since the frequency of an ion acoustic wave is much less than ω_0, it is clear that this instability can occur throughout the underdense plasma. Furthermore, nearly all the energy can be transferred to the scattered light wave. Hence this instability is a significant concern for laser fusion applications, since the process can either degrade the absorption or change its location.

8.1   INSTABILITY ANALYSIS

To obtain the coupled equations [1,2] describing the Brillouin instability, we again consider the response of an initially uniform plasma driven by a large amplitude light wave. We have already derived in the previous chapter an equation for the generation of a scattered light wave with vector potential $\tilde{\mathbf{A}}$ by the coupling of a large amplitude light wave with vector potential $\mathbf{A}_L$ with an electron density fluctuation $\tilde{n}_e$:

$$\left(\frac{\partial^2}{\partial t^2} - c^2 \nabla^2 + \omega_{\mathrm{pe}}^2 \right) \tilde{\mathbf{A}} = - \frac{4\pi \, e^2}{m} \, \tilde{n}_e \, \mathbf{A}_L \, , \qquad (8.1)$$

where ω_{pe} is the electron plasma frequency. Only the fluctuation in electron density appears in Eq. (8.1), since the ion resonse to the high frequency field of the light wave is less than the electron response by Zm/M, where Z is the charge state, m the electron mass, and M the ion mass.

For the Brillouin instability, the density fluctuation $\tilde{n}_e$ is the low frequency fluctuation associated with an ion acoustic wave. To derive an equation for this low frequency fluctuation, the ion motion must also be included. We again describe the electrons as a warm fluid and separate the fluid velocity $\mathbf{u}_e$ into longitudinal ($\mathbf{u}_L$) and transverse components ($e\mathbf{A}/mc$). Then, as shown in Eq. (7.14),

$$\frac{\partial \mathbf{u}_L}{\partial t} = \frac{e}{m} \nabla \phi - \frac{1}{2} \nabla \left(\mathbf{u}_L + \frac{e\mathbf{A}}{mc} \right)^2 - \frac{\nabla p_e}{n_e m} \, , \qquad (8.2)$$

where ϕ is the electrostatic potential, p_e the electron pressure, and n_e the electron density. Since we are now considering a low frequency fluctuation, we neglect the electron inertia ($\partial \mathbf{u}_L/\partial t \to 0$) and use the isothermal equation of state ($p_e = n_e \theta_e$, where θ_e is the electron temperature). We then linearize Eq. (8.2) by letting $n_e = n_0 + \tilde{n}_e$, $\mathbf{A} = \mathbf{A}_L + \tilde{\mathbf{A}}$ and $\phi = \tilde{\phi}$, which gives

$$\frac{e}{m}\nabla\tilde{\phi} = \frac{e^2}{m^2 c^2}\nabla\left(\mathbf{A}_L \cdot \tilde{\mathbf{A}}\right) + \frac{v_e^2}{n_0}\nabla\tilde{n}_e \,, \qquad (8.3)$$

where v_e is the electron thermal velocity. The electrical potential transmits the ponderomotive force to the ions.

To treat the ion response, we describe the ions as a charged fluid with density n_i and velocity $\mathbf{u}_i$. The continuity and force equations are

$$\frac{\partial n_i}{\partial t} + \nabla \cdot (n_i \mathbf{u}_i) = 0$$
$$\frac{\partial \mathbf{u}_i}{\partial t} + \mathbf{u}_i \cdot \nabla \mathbf{u}_i = -\frac{Z\,e}{M}\nabla\phi \,, \qquad (8.4)$$

where we have neglected the ion pressure for simplicity. We next linearize these equations by taking $n_i = n_{0i} + \tilde{n}_i$, $\mathbf{u}_i = \tilde{\mathbf{u}}_i$ and $\phi = \tilde{\phi}$. Then

$$\frac{\partial \tilde{n}_i}{\partial t} + n_{0i}\nabla \cdot \tilde{\mathbf{u}}_i = 0 \qquad (8.5)$$

$$\frac{\partial \tilde{\mathbf{u}}_i}{\partial t} = -\frac{Z\,e}{M}\nabla\tilde{\phi} \,. \qquad (8.6)$$

Taking a time derivative of Eq. (8.5), a divergence of Eq. (8.6) and combining to eliminate the term $\partial\nabla \cdot \tilde{\mathbf{u}}_i/\partial t$ gives

$$\frac{\partial^2 \tilde{n}_i}{\partial t^2} - \frac{n_{0i}Ze}{M}\nabla^2\tilde{\phi} = 0 \,. \qquad (8.7)$$

If we substitute for $\tilde{\phi}$ using Eq. (8.3), note that $Zn_{0i} = n_0$ and approximate $Z\tilde{n}_i \simeq \tilde{n}_e$, we finally obtain an equation for the low frequency density fluctuation:

$$\frac{\partial^2 \tilde{n}_e}{\partial t^2} - c_s^2 \nabla^2\tilde{n}_e = \frac{Z\,n_0 e^2}{m\,M\,c^2}\nabla^2(\mathbf{A}_L \cdot \tilde{\mathbf{A}}) \,. \qquad (8.8)$$

Here $c_s = (Z\theta_e/M)^{1/2}$ is the ion acoustic velocity. Equation (8.8) describes the excitation of an ion acoustic wave by the interaction between the incident and scattered light waves.

8.2   DISPERSION RELATION

To derive the dispersion relation from the coupled equations for $\tilde{\mathbf{A}}$ and $\tilde{n}_e$, we take $\mathbf{A}_L = A_L \cos(\mathbf{k}_0 \cdot \mathbf{x} - \omega_0 t)$ and Fourier-analyze Eqs. (8.1) and (8.8) to obtain

$$D(k,\omega)\,\tilde{\mathbf{A}}(k,\omega) = \frac{4\pi e^2}{m}\frac{\mathbf{A}_L}{2}\left[\tilde{n}_e(k-k_0,\omega-\omega_0)+\tilde{n}_e(k+k_0,\omega+\omega_0)\right] \quad (8.9)$$

$$(\omega^2 - k^2 c_s^2)\,\tilde{n}_e(k,\omega) = \qquad\qquad\qquad\qquad\qquad (8.10)$$
$$\frac{Zn_0 e^2}{mMc^2}\frac{k^2 \mathbf{A}_L}{2}\cdot\left[\tilde{\mathbf{A}}(k-k_0,\omega-\omega_0)+\tilde{\mathbf{A}}(k+k_0,\omega+\omega_0)\right],$$

where $D(k,\omega) = \omega^2 - k^2 c^2 - \omega_{\text{pe}}^2$. We next use Eq. (8.9) to eliminate $\tilde{\mathbf{A}}(k-k_0,\omega-\omega_0)$ and $\tilde{\mathbf{A}}(k+k_0,\omega+\omega_0)$ from Eq. (8.10). If we choose ω to be low frequency ($\omega \ll \omega_0$) and neglect as nonresonant the terms with $n_e(k\pm 2k_0,\omega\pm 2\omega_0)$, we obtain the dispersion relation:

$$\omega^2 - k^2 c_s^2 = \frac{k^2 v_{\text{os}}^2}{4}\omega_{\text{pi}}^2\left[\frac{1}{D(\omega-\omega_0,\mathbf{k}-\mathbf{k}_0)}+\frac{1}{D(\omega+\omega_0,\mathbf{k}+\mathbf{k}_0)}\right].(8.11)$$

Here $v_{\text{os}} = eA_L/mc$, and ω_{pi} is the ion plasma frequency which is given by $\omega_{\text{pi}} = \omega_{\text{pe}}\sqrt{Zm/M}$.

Instability growth rates are readily found from Eq. (8.11). For Brillouin back or sideward scatter, k is of order k_0 and so only the downshifted light wave need be retained. Then

$$\left(\omega^2 - k^2 c_s^2\right)\left(\omega^2 - 2\omega\omega_0 + 2\mathbf{k}_0\cdot\mathbf{k}c^2 - k^2 c^2\right) = \frac{k^2 v_{\text{os}}^2}{4}\omega_{\text{pi}}^2. \quad (8.12)$$

As an example, we consider backscatter which has the largest growth rate. If we consider first the weak field limit in which $\omega = kc_s + i\gamma$, where $\gamma \ll kc_s$, Eq. (8.12) becomes

$$2ikc_s\gamma\left(-2i\omega_0\gamma - 2\omega_0 kc_s + 2kk_0 c^2 - k^2 c^2\right) = \frac{k^2 v_{\text{os}}^2}{4}\omega_{\text{pi}}^2. \quad (8.13)$$

Maximum growth clearly occurs for k such that the scattered light wave is also a resonant mode. Then

$$k = 2k_0 - \frac{2\omega_0}{c}\frac{c_s}{c} \qquad\qquad\qquad (8.14)$$

$$\gamma = \frac{1}{2\sqrt{2}}\frac{k_0 v_{\text{os}}\,\omega_{\text{pi}}}{\sqrt{\omega_0 k_0 c_s}}. \qquad\qquad (8.15)$$

In the strong field limit, $|\omega| \gg k c_s$, and Eq. (8.11) then becomes a cubic equation for ω. Again choosing k as given by Eq. (8.14) for maximum growth, we now obtain

$$\omega \simeq \left(\frac{k_0^2 \, v_{os}^2}{2} \frac{\omega_{pi}^2}{\omega_0} \right)^{1/3} \left[\frac{1}{2} + i \, \frac{\sqrt{3}}{2} \right]. \qquad (8.16)$$

Note that in this strong field limit, the frequency of the electrostatic wave is determined by the amplitude of the light wave. In this limit, the electrostatic wave is sometimes called a quasi-mode, since it is not a normal mode of an undriven plasma.

The wave number k corresponding to sidescatter is less than that for backscatter, since the ion wave has to take up less momentum. For example, for 90° sidescatter, $k = \sqrt{2} \, k_0$. As is then apparent from Eq. (8.11), the growth rate is also less for sidescatter. We again note that sidescatter occurs preferentially into light waves propagating out of the plane defined by the electric and propagation vectors of the large amplitude light wave. In this case, the electric vectors of the light waves can be aligned, maximizing the ponderomotive force. This is the case we have focused on with our simplifying assumption that $\mathbf{A} \cdot \nabla n_e = 0$.

There is also a kinetic version of the Brillouin instability, which represents stimulated scattering from the ions. In this instability, the two electromagnetic waves beat together to produce an electrostatic fluctuation which resonates with the bulk of the ions i.e., $\omega_0 - \omega_s \sim k v_i$, where v_i is the ion thermal velocity. Since the electrostatic disturbance is not a normal mode of the plasma, the growth rate is much less than that for the Brillouin instability unless the ion waves are heavily damped. This instability can be included in the dispersion relation by replacing our fluid description of the ions with a kinetic treatment.

8.3 INSTABILITY THRESHOLDS

Damping of the unstable waves introduces a threshold intensity for instability generation. As we have discussed in the previous chapter, net growth of the unstable pair of waves requires that

$$\gamma > \sqrt{\gamma_i \gamma_s}, \qquad (8.17)$$

where γ is the growth rate in the absence of damping, γ_s is the amplitude damping rate of the scattered light wave, and γ_i is the amplitude damping rate of the ion acoustic wave. As an example, let us consider backscatter and assume collisional damping of the light wave ($\gamma_s \simeq \nu_{ei}\omega_{\mathrm{pe}}^2/2\omega_0^2$, where ν_{ei} is the electron-ion collision frequency defined in Chapter 5). Substituting Eq. (8.15) into Eq. (8.17) then gives

$$\left(\frac{v_{os}}{v_e}\right)^2 \geq 4\,\frac{\nu_{ei}}{\omega_0}\frac{\gamma_i}{k_0\,c_s}\,. \tag{8.18}$$

The damping of the ion wave is usually determined by Landau damping, which will be discussed in a later chapter. We need only note here that $\gamma_i < k_0 c_s$, and so the threshold due to damping is usually quite low i.e., $(v_{os}/v_e)^2 \ll 1$.

In practice, the threshold intensity is usually determined by gradients in the plasma density and expansion velocity rather than by damping. As indicated in the previous chapter, instability growth in an inhomogeneous plasma is a very rich topic. To illustrate the effects of inhomogeneity, let's again simply consider the Rosenbluth criterion for $\exp(2\pi)$ convective growth in a plasma with a linear variation in the wavenumber matching:

$$\frac{\gamma^2}{|\kappa'\,v_{g1}\,v_{g2}|} \gtrsim 1\,. \tag{8.19}$$

Here κ' is the gradient of the wave number mismatch and v_{g1} and v_{g2} are the components of the group velocities of the unstable waves along this gradient. As an example, we consider Brillouin backscatter in a plasma with a density gradient with scale length $L = n/(\partial n/\partial x)$. Then $\kappa' = \partial(k_0 - k_s - k)/\partial x \simeq 2(\partial k_0/\partial x) \simeq \omega_{\mathrm{pe}}^2/\omega_0 cL$, for $\omega_{\mathrm{pe}} \ll \omega_0$. Substituting into Eq (8.19) gives

$$\left(\frac{v_{os}}{v_e}\right)^2 \gtrsim \frac{8}{k_0\,L}\,. \tag{8.20}$$

A gradient in expansion velocity can be even more effective in limiting the region over which the coupling is resonant. In an expanding plasma, $\omega = k(c_s + v_{\exp})$. Then $\kappa' = -\partial k/\partial x \simeq kc_s L_v^{-1}/(c_s + v_{\exp})$, where $L_v = c_s/(\partial v_{\exp}/\partial x)$ is the velocity gradient scale length. Substituting κ' into Eq. (8.19), we then obtain the result shown in Eq. (8.20) with the replacement of L by $L_v\,\omega_{\mathrm{pe}}^2/2\omega_0^2$.

The threshold intensity due to inhomogeneity is lower for sidescatter, since a sidescattered light wave spends a longer time in the region of interaction. If we refer to the discussion in Chapter 3, the group velocity along the density gradient of a light wave propagating principally in the direction orthogonal to the gradient is $\sim c/(k_0 L)^{1/3}$. As is apparent from Eq. (8.19), the threshold intensity is then reduced by a factor of order $(k_0 L)^{1/3}$.

8.4   THE FILAMENTATION INSTABILITY

The dispersion relation shown in Eq. (8.11) also describes the filamentation instability [3–6], which corresponds to the growth of zero-frequency density perturbations (and the corresponding modulations in intensity) in the plane orthogonal to the propagation vector of the light wave. If we assume that $\omega = i\gamma \ll \omega_0$ and $\mathbf{k} \cdot \mathbf{k}_0 = 0$, we find that $D(\omega \pm \omega_0, \mathbf{k} \pm \mathbf{k}_0) \simeq \pm 2i\omega_0\gamma - k^2 c^2$. Substituting into Eq. (8.11), we obtain

$$(\gamma^2 + k^2 c_s^2)\left(\gamma^2 + \frac{k^4 c^4}{4\omega_0^2}\right) = \frac{k^4 v_{os}^2 c^2}{8} \frac{\omega_{pi}^2}{\omega_0^2}. \tag{8.21}$$

For illustration, let us here derive the maximum growth rate in the limit $\gamma \ll kc_s$. The wave number for maximum growth is found by differentiating Eq. (8.21) with respect to k and requiring that $\partial\gamma/\partial k = 0$. The growth rate is then evaluated for this value of k. In the limit $\gamma \ll kc_s$ and neglecting the ion temperature, we obtain

$$\gamma = \frac{1}{8}\left(\frac{v_{os}}{v_e}\right)^2 \frac{\omega_{pe}^2}{\omega_0}$$
$$k = \frac{\omega_{pe}}{2c}\frac{v_{os}}{v_e}.$$

We note that the density fluctuations are purely growing. They simply correspond to the variations in plasma density driven via the ponderomotive force by intensity modulations in the light beam. Whole beam self-focusing results from the same physical process and can be considered a special case of the filamentation process. Since resonance with an ion wave is not involved, the filamentation instability is not extremely sensitive to plasma inhomogeneity. The instability is often characterized by its spatial gain coefficient κ, which is the growth rate divided by the group velocity of the light wave. For $\gamma \ll kc_s$, $\kappa \simeq (1/8)(v_{os}/v_e)^2 (\omega_{pe}^2/\omega_0^2)(\omega_0/c)$.

Filamentation and self-focusing can also be driven by either thermal forces [7–10] or relativistic effects [11]. In the first case, a localized increase in the intensity of a light wave raises the plasma temperature via the enhanced heating. Refraction of the light wave into the resulting density depression enhances the perturbation in intensity, completing the feedback loop. These thermal effects can be particularly important in dense, cold plasmas where collisional absorption is efficient. The relativistic effect can be significant for a very intense light wave. Since $\omega_{pe}^2 = 4\pi n e^2/m$, the relativistic increase in the mass of an electron oscillating in the light wave has the same effect as a decrease in the plasma density. The light wave is focused, enhancing its intensity.

References

1. Drake, J. F., P. K. Kaw, Y. C. Lee, G. Schmidt, C. S. Liu, and M. N. Rosenbluth, Parametric instabilities of electromagnetic waves in plasmas, *Phys. Fluids* **17**, 778 (1974).

2. Liu, C. S., Parametric instability in an inhomogeneous unmagnetized plasma; in *Advances in Plasma Physics*, Vol. 16, (A. Simon and W. Thompson, eds.), p. 121–177. Wiley, New York, 1976.

3. Kaw, P. K., G. Schmidt and T. Wilcox, Filamentation and trapping of electromagnetic radiation in plasmas, *Phys. Fluids* **16**, 1522 (1973).

4. Litvak, A. G., Finite-amplitude wave beams in a magnetoactive plasma, *Sov Phys. —JETP* **30**, 344 (1970).

5. Palmer, A. J., Stimulated scattering and self-focusing in laser-produced plasmas, *Phys. Fluids* **14**, 2714 (1971).

6. Cohen, B. I. and C. E. Max, Stimulated scattering of light by ion modes in a homogeneous plasma: space-time evolution, *Phys. Fluids* **22**, 1115 (1979).

7. Sodka, M. S., A. K. Ghatak and V. K. Tripathi, Self-focusing of laser beams in plasmas and semiconductors; in *Progress in Optics*, Vol. 13, (E. Wolk, ed.). North-Holland, Amsterdam, 1976.

8. Perkins, F. and E. Valeo, Thermal self-focusing of electromagnetic waves in plasmas, *Phys. Rev Letters* **32**, 1234 (1974).

9. Craxton, R. S. and R. L. McCrory, Hydrodynamics of thermal self-focusing in laser plasmas, *J. Appl. Phys.* **56**, 108 (1984).

10. Estabrook, K. and W. L. Kruer, Two-dimensional ray-trace calculations of thermal whole beam self-focusing, *Phys. Fluids* **28**, 19 (1985).

11. Max, C. E., J. Aarons and A. B. Langdon, Self-modulation and self-focusing of electromagnetic waves in plasmas, *Phys. Rev. Letters* **33**, 209 (1974).

CHAPTER 9

Heating by Plasma Waves

The Wave-Particle Interaction

We have examined a number of different processes whereby intense laser light couples either into electrostatic waves (resonance absorption and the oscillating-two-stream, ion acoustic decay and $2\omega_{pe}$ instabilities) or into both electrostatic and scattered light waves (the Raman and Brillouin instabilities). In order to understand the evolution and the consequences of these processes, it is necessary to consider how electrostatic waves are damped by the plasma particles. Since electrostatic waves are simply charge density fluctuations and their associated electric fields, these waves do not readily escape from the plasma. Their energy is ultimately transferred to the particles via either linear or nonlinear damping mechanisms. We will first discuss the damping of a small amplitude electron plasma wave, which is already sufficient to illustrate important features of the heating via plasma waves. We will then briefly consider the damping of a large amplitude electron plasma wave. Finally we conclude with a discussion of electron heating by parametric instabilities near the critical density, including a brief discussion of plasma wave collapse.

9.1 COLLISIONAL DAMPING

Electron-ion collisions provide the simplest mechanism for the damping of an electron plasma wave. Our discussion of this collisional damping is quite analogous to that previously given for an electromagnetic wave. The coherent motion of oscillation of electrons in the electric field of the wave is converted to random (or thermal) motion at the rate at which electron-ion collisions occur. To balance the energy dissipated, the energy of the wave then damps at the rate ν i.e.,

$$\frac{\nu E^2}{8\pi} = \nu_{ei} \frac{n m v_w^2}{2}, \qquad (9.1)$$

where $v_w = eE/m\omega$, E is the amplitude and ω is the frequency of the electric field, n is the plasma density, and ν_{ei} is the electron-ion collision frequency. Hence $\nu = \omega_{pe}^2 \nu_{ei}/\omega^2$, where ω_{pe} is the electron plasma frequency. The value of ν_{ei} is the same as that derived in Chapter 5 for the collisional damping of a light wave. For an electron plasma wave, $\omega \simeq \omega_{pe}$ and so $\nu \simeq \nu_{ei}$.

As discussed in previous chapters, collisional damping can determine the threshold intensity for instabilities, an effect which can be quite significant for dense, low temperature, high Z plasmas. However, the collisional thresholds are often greatly exceeded, particularly in laser-fusion applications with very intense and/or long wavelength laser light. Other mechanisms for the wave damping must then be considered.

9.2 LANDAU DAMPING

An electrostatic wave can be damped even in the absence of collisions. This so-called collisionless or Landau damping can be qualitatively understood rather simply. Consider an electrostatic wave with electric field $E \sin(kx - \omega t)$. Most particles are non-resonant i.e., have a velocity v much different than ω/k, the phase velocity of the wave. These particles simply oscillate in the field and experience no secular gain or loss in energy. In contrast, resonant particles with $v \simeq \omega/k$ experience a nearly constant field and so can be efficiently accelerated or decelerated. These particles do exchange energy with the wave.

A very straight forward and physical treatment of Landau damping can be given [1]. We will first calculate the changes in the energy of particles moving in a given field. Then we will average these energy changes over a distribution of particles. Finally we will invoke energy balance to determine the rate at which the field damps or grows due to its interaction with the particles.

Since we consider an electrostatic wave and neglect any magnetic fields, a one dimensional treatment is sufficient. The particle dynamics are determined by

$$\ddot{x} = \frac{q}{m} E \sin(kx - \omega t) , \qquad (9.2)$$

where q and m are the charge and mass and E is the amplitude of the electric field. We compute the dynamics by expanding about the free-streaming motion of a particle with initial position x_0 and initial velocity v_0. In particular, we assume that

$$x = x_0 + v_0 t + x_1 + x_2 \qquad (9.3)$$
$$v = v_0 + v_1 + v_2 . \qquad (9.4)$$

The subscript 1 denotes a first-order correction which is proportional to E; the subscript 2 denotes a second-order correction proportional to E^2. As will become apparent, our expansion parameter is $k\delta x$, where δx is the change between the free-streaming position of a particle and its actual position.

To compute the energy changes to order E^2, we must simply compute the motion to second order. If we substitute Eqs. (9.3) and (9.4) into Eq. (9.2) and expand, we obtain

$$\dot{v}_1 = \frac{qE}{m} \sin(kx_0 - \Omega t) \qquad (9.5)$$

$$\dot{v}_2 = \frac{qE}{m} kx_1 \cos(kx_0 - \Omega t) , \qquad (9.6)$$

where $\Omega = \omega - kv_0$. Several trivial integrations of Eq. (9.5) give

$$v_1 = \frac{qE}{m\Omega} \left[\cos(kx_0 - \Omega t) - \cos kx_0 \right] \qquad (9.7)$$

$$x_1 = -\frac{qE}{m\Omega^2} \left[\sin(kx_0 - \Omega t) - \sin kx_0 + \Omega t \cos kx_0 \right] . \qquad (9.8)$$

Substitution of Eq. (9.8) into Eq. (9.6) gives

$$\dot{v}_2 = -\frac{kq^2 E^2}{m^2\Omega^2} \cos(kx_0 - \Omega t) \left[\sin(kx_0 - \Omega t) - \sin kx_0 + \Omega t \cos kx_0 \right]. \qquad (9.9)$$

We next compute the rate of change of the energy $(\delta\dot{\mathcal{E}})$ of a set of particles with random initial positions. First, note that $\langle\delta\dot{\mathcal{E}}_1\rangle = \langle mv_0\dot{v}_1\rangle = 0$, where $\langle\ \rangle$ denotes the average over initial positions. To second order, we obtain $\langle\delta\dot{\mathcal{E}}_2\rangle = mv_0\langle\dot{v}_2\rangle + m\langle v_1\dot{v}_1\rangle$. Substituting from Eqs. (9.7) and (9.9) gives

$$\langle\delta\dot{\mathcal{E}}_2\rangle = \frac{q^2E^2}{2\,m}\left[\frac{\sin\Omega t}{\Omega} + \frac{kv_0}{\Omega^2}\left(\sin\Omega t - \Omega t\cos\Omega t\right)\right]. \qquad (9.10)$$

Considerable simplification results if we now take the long-time limit and express the results in terms of a delta function. A useful representation of a delta function is

$$\delta(\Omega) = \lim_{t\to\infty}\frac{\sin\Omega t}{\pi\,\Omega}. \qquad (9.11)$$

Hence Eq. (9.10) can be expressed as

$$\langle\delta\dot{\mathcal{E}}_2\rangle = \frac{\pi\,q^2E^2}{2\,m}\left[\delta(\Omega) - kv_0\frac{\partial}{\partial\Omega}\delta(\Omega)\right]. \qquad (9.12)$$

Since $\delta(\omega - kv_0) = |k|^{-1}\,\delta(v_0 - \omega/k)$, we then obtain

$$\langle\delta\dot{\mathcal{E}}_2\rangle = \frac{\pi q^2E^2}{2\,m\,|k|}\frac{\partial}{\partial v_0}\left[v_0\,\delta(v_0 - \frac{\omega}{k})\right]. \qquad (9.13)$$

Lastly, we average the rate of the energy change over a distribution of initial velocities, $f(v_0)$. Then

$$\overline{\langle\delta\dot{\mathcal{E}}_2\rangle} = \int\,dv_0\,f(v_0)\,\langle\delta\dot{\mathcal{E}}_2\rangle, \qquad (9.14)$$

where the bar denotes the average over velocities. Substituting Eq. (9.13) into Eq. (9.14) and integrating gives the rate at which the particles gain or lose energy:

$$\overline{\langle\delta\dot{\mathcal{E}}_2\rangle} = -\frac{\pi\,q^2E^2}{2\,m\,|k|}\frac{\omega}{k}\frac{\partial f}{\partial v}\left(\frac{\omega}{k}\right). \qquad (9.15)$$

Equation (9.15) illustrates some very important features of the wave-particle interaction in the collisionless limit. The energy exchange is determined by the resonant particles (those with $v_0 \simeq \omega/k$) and depends on the slope of the velocity distribution at the phase velocity of the wave. In particular, particles with velocity slightly less than ω/k gain energy;

particles with velocity slightly greater than ω/k lose energy. If the velocity distribution decreases with velocity, the particles gain energy from the wave. If the slope of the distribution function is inverted, the particles lose energy to the wave.

By energy conservation, the rate of change in the energy of the particles must be balanced by a growth or damping of the wave. Specializing to an electron plasma wave, we have

$$2\gamma\,\frac{E^2}{8\pi} + \overline{\langle\delta\dot{\mathcal{E}}_2\rangle} = 0\,, \tag{9.16}$$

where γ is the rate at which the electric field grows or damps. Substituting from Eq. (9.15) gives

$$\frac{\gamma}{\omega} = \frac{\pi}{2}\,\frac{\omega_{\mathrm{pe}}^2}{k^2}\,\frac{\partial}{\partial v}\bar{f}\!\left(\frac{\omega}{k}\right)\,, \tag{9.17}$$

where $f = n\bar{f}$ and ω_{pe} is the electron plasma frequency with density n. Note that γ depends on the slope of the distribution function evaluated at the phase velocity of the wave. This Landau damping (or growth) rate can also be readily derived directly from the Vlasov equation.

If we consider as an example a Maxwellian distribution with thermal velocity v_e,

$$\frac{\gamma}{\omega} = -\sqrt{\frac{\pi}{8}}\,\frac{\omega_{\mathrm{pe}}^2\,\omega}{|k^3|\,v_e^3}\,\exp\!\left(-\frac{\omega^2}{2k^2v_e^2}\right)\,. \tag{9.18}$$

Note that the Landau damping of an electron plasma wave is a strong function of its phase velocity. The damping becomes sizeable whenever $\omega/k \lesssim 3v_e$ i.e., when $k\lambda_{\mathrm{De}} \gtrsim 0.4$ where λ_{De} is the electron Debye length and $\omega = (\omega_{\mathrm{pe}}^2 + 3k^2v_e^2)^{1/2}$.

Let us conclude our discussion of linear Landau damping with a simple mechanical analogy. Consider a group of boxes translating along at a velocity equal to ω/k. Inside the boxes are uniformly-distributed particles, some moving slightly slower than ω/k, some moving slightly faster. As illustrated in Fig. 9.1, those particles moving slower than ω/k are overtaken by the wall to their left and gain energy as they are bounced off. Likewise, those particles moving faster than ω/k overtake the right wall and lose energy as they are reflected. For a time less than the transit time of a particle through the box, the net energy change simply depends on whether more particles are initially moving faster or slower than ω/k.

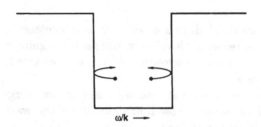

Figure 9.1   A mechanical analogue for Landau damping.

9.3   LINEAR THEORY LIMITATIONS — TRAPPING

Let us now note some important restrictions on the linear theory of the wave-particle interaction. Our expansion about the free-streaming orbit requires that $k\delta x \ll 1$, a condition that fails after a finite time which depends on the wave amplitude. This limitation on the linear theory can easily be seen by examining $\langle k^2 x_1^2 \rangle$. Using Eq. (9.8), we obtain

$$\langle k^2 x_1^2 \rangle = \frac{k^2 e^2 E^2}{2\,m^2 \Omega^4} \left[(1 - \cos \Omega t)^2 + (\Omega t - \sin \Omega t)^2 \right].$$

The condition $\langle k^2 x_1^2 \rangle \ll 1$ is most stringent for a resonant particle ($\Omega = 0$): $(k^2 e^2 E^2/8m^2)\,t^4 \ll 1$. Defining a frequency $\omega_b = (eEk/m)^{1/2}$, we then have $\omega_b t \ll 8^{1/4}$.

Physically, ω_b is a characteristic frequency with which trapped electrons oscillate in the potential troughs of the wave. Consider the motion of a resonant electron in the field $E \sin(kx - \omega t)$. If we use the transformation $\xi = x - (\omega/k)t$ to change to a frame moving with the phase velocity, Eq. (9.2) becomes

$$\ddot{\xi} = -\frac{e}{m} E \sin k\xi . \tag{9.19}$$

For electrons near the bottom of the potential troughs (i.e., well trapped electrons), $\sin k\xi \simeq k\xi$. Hence,

$$\ddot{\xi} = -\frac{e k E}{m}\xi , \tag{9.20}$$

describing harmonic motion with a bounce frequency $\omega_b = (eEk/m)^{1/2}$. The linear theory only describes the early phase of this motion i.e., for

$\omega_b t \ll 1$. Alternatively, the linear theory requires that $\gamma \gg \omega_b$ i.e., that the wave damp before electrons can oscillate in the troughs.

In the opposite limit ($\gamma \ll \omega_b$), we encounter trapping of electrons in the potential troughs of the electrostatic wave. Since the motion of the resonant particles (both trapped and untrapped) becomes periodic, we then expect the amplitude of the wave to oscillate, as it first gives and then recovers energy from the particles. In other words, first there are more electrons moving slightly slower than the phase velocity of the wave. The wave damps as the slower electrons gain energy. This leads to a situation in which there are now more electrons moving slightly faster than ω/k, and the wave regains energy from the particles. Such an oscillation in the energy exchange is also present in the mechanical analogy discussed in the previous section. Electrons bouncing off one wall and gaining energy clearly lose this energy after they transit the box and bounce off the other wall.

Of course, electrons in a sinusoidal potential trough actually have bounce frequencies which depend on their initial positions. Hence the periodic interchange of energy between the wave and the particles gradually phase mixes away, as the slope of the distribution function flattens in the neighborhood of the phase velocity [2]. It should also be noted that the trapped electrons can also generate the so-called sideband instability, which leads to an exponentiation of nearby waves [3].

9.4 WAVEBREAKING OF ELECTRON PLASMA WAVES

Let us now discuss a useful picture which illustrates some important qualitative features of the nonlinear wave-particle interaction. As linear theory has shown, a small amplitude wave is damped only by those particles with a velocity quite near its phase velocity. However, in a large amplitude electron plasma wave, the oscillation velocity of an electron in the field can be large enough to bring even an initially cold, main body particle into resonance with the field. That is, when $(eE/m\omega) \simeq \omega/k$, numerous particles can "resonantly interact" with the wave. A strong, nonlinear damping results as electrons are efficiently accelerated by the wave. The wave amplitude is often referred to as the amplitude at which breaking occurs in a cold plasma [4].

At the wave breaking amplitude, large numbers of formerly nonresonant particles become strongly "trapped." The wave energy is suddenly damped as these slow particles are accelerated by falling into the potential troughs of the wave. At this wavebreaking, $\omega_b = (eEk/m)^{1/2} = \omega$. Since ω_b^{-1} is the characteristic time for resonant particles to move in the field and hence take energy from it, the energy exchange to the particles takes place very rapidly.

The amplitude of the field at which particles are nonlinearly brought into resonance with a wave (i.e., are strongly trapped) is significantly reduced [5] in a warm plasma for several reasons. Faster electrons are more easily brought into resonance, and the sizeable pressure force associated with the density fluctuation of the wave gives an additional acceleration.

We can crudely model the effect of plasma temperature on wavebreaking by considering a water bag model, which corresponds to replacing a Maxwellian distribution with a velocity distribution which is constant between $\pm\sqrt{3}v_e$. Such an idealized distribution is convenient since it has the same pressure as does a Maxwellian distribution with thermal velocity v_e, yet there is a well-defined maximum initial velocity of $\sqrt{3}\,v_e$. Although there are particles with an arbitrarily high velocity present in a Maxwellian distribution, the number of particles is not sizeable until $v \lesssim 2v_e$. Hence the water bag distribution can be expected to roughly model the condition that significant numbers of particles are nonlinearly brought into resonance.

In this model which assumes fixed ions, the average density (n) and velocity (u) satisfy the same equations as those for a warm electron fluid, as is apparent from taking moments of the Vlasov equation. Hence the continuity and force equations are

$$\frac{\partial n}{\partial t} + \frac{\partial}{\partial x}(nu) = 0\,, \tag{9.21}$$

$$\frac{\partial u}{\partial t} + u\frac{\partial u}{\partial x} = -\frac{e}{m}E - \frac{1}{mn}\frac{\partial p}{\partial x}\,. \tag{9.22}$$

Since we are considering a high frequency electron plasma wave, the pressure p is determined by the adiabatic equation of state. Introducing $E = -\partial\phi/\partial x$ and transforming to the wave frame with velocity $-v_p$ gives

$$nu = n_0 v_p \tag{9.23}$$

$$u^2 - \frac{2e\phi}{m} + 3v_e^2\frac{n^2}{n_0^2} = v_p^2 + 3v_e^2\,. \tag{9.24}$$

Here v_e is the electron thermal velocity and n_0 is the density of the uniform, unperturbed plasma. Substituting Eq. (9.23) into Eq. (9.24), we obtain

$$\frac{2\,e\,\phi}{m\,v_p^2} = \frac{u^2}{v_p^2} - 1 - \beta + \beta\,\frac{v_p^2}{u^2}, \tag{9.25}$$

where $\beta = 3v_e^2/v_p^2$. By differentiating Eq. (9.25) with respect to u, it is easy to see that ϕ has an extremum (ϕ_{cr}) when $u/v_p = \beta^{1/4}$. The corresponding potential is

$$-\frac{2\,e\,\phi_{cr}}{m\,v_p^2} = \left(1 - \sqrt{\beta}\right)^2. \tag{9.26}$$

This simply corresponds to the condition that the net energy of the fastest electron be zero in the wave frame.

To determine the critical value of the electric field, we consider Poisson's equation: $\partial^2\phi/\partial x^2 = 4\pi e\,(n - n_0)$. Multiplying by $\partial\phi/\partial x$ and using Eqs. (9.23) and (9.24) gives (in the wave frame)

$$\frac{\dot{\phi}^2}{2} + 4\pi\left[n_0\,e\phi - nmu^2 - n_0 m\,v_e^2\,\frac{n^3}{n_0^3}\right] = \tag{9.27}$$

$$-2\pi\,n_0\,m\,v_p^2\left[\left(1 - \sqrt{\beta}\right)^2 + \frac{8}{3}\beta^{1/4}\right].$$

The constant has been evaluated by noting that $\dot{\phi} = 0$ when $\phi = \phi_{cr}$. The maximum electric field ($E_{max} = -\dot{\phi}_{max}$) obtains when $\phi = 0$:

$$\frac{e^2\,E_{max}^2}{m^2\,\omega_{pe}^2\,v_p^2} = 1 + 2\sqrt{\beta} - \frac{8}{3}\beta^{1/4} - \frac{\beta}{3}. \tag{9.28}$$

The maximum field amplitude is plotted in Fig. 9.2 as a function of $\sqrt{3}\,v_e/v_p$. For $v_e = 0$, the cold plasma result is recovered. Note the sizeable decrease of the maximum field as the plasma temperature increases. For example, for $v_p = 5v_e$, $(eE_{max}/m\,\omega_{pe}\,v_p) \simeq 0.3$.

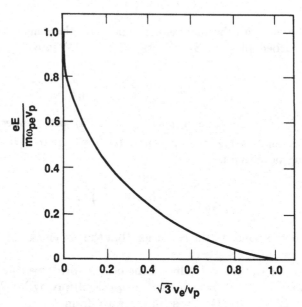

Figure 9.2   The wavebreaking amplitude as a function of thermal velocity.

9.5 ELECTRON HEATING BY THE OSCILLATING-TWO-STREAM AND ION ACOUSTIC DECAY INSTABILITIES

We will continue our consideration of electron heating via plasma waves with a discussion of some particle simulations of a simple but instructive model problem. A plasma with a uniform density is driven by an imposed spatially-independent pump field ($E_0 \sin \omega_0 t$) with a frequency (ω_0) near the electron plasma frequency. Such a pump field models the electric field of a light wave near its critical density under the assumption that the wave number of the light wave is negligible compared with the wave numbers of the plasma waves which are excited. Since the unstable plasma waves preferentially grow along the electric vector of the pump field, a great deal can be learned by using a one-dimensional electrostatic particle code [6–8].

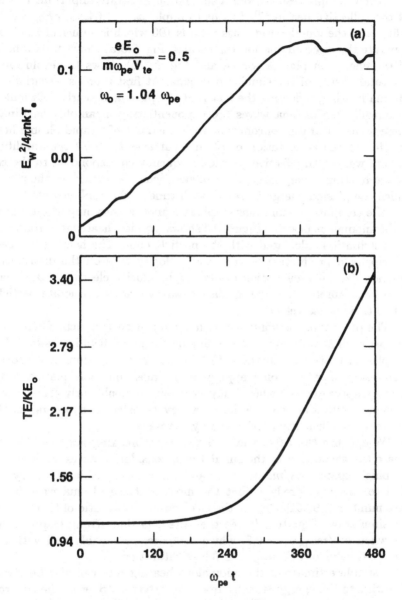

Figure 9.3   Computed evolution of (a) the electron plasma wave energy and (b) the total energy of a plasma driven by an electric field oscillating near the electron plasma frequency (from *Kruer et al.*, 1970).

A few sample results from a simulation illustrate important features of the collective heating[7]. In this example $\omega_0 = 1.04\omega_{\mathrm{pe}}$, $eE_0/m\omega_{\mathrm{pe}} = 0.5v_e$, and the ion-electron mass ratio is 100 which is sufficient to clearly separate the electron and ion time scales. Fig. 9.3(a) shows the evolution of the energy in plasma waves, and Fig. 9.3(b) shows the evolution of the total energy of the simulated plasma. At first there is essentially no plasma heating, reflecting the fact that the plasma is nearly collisionless. Meanwhile the plasma waves are exponentiating in amplitude. Finally these waves saturate, concomitant with the onset of a rapid plasma heating due to the acceleration of plasma particles by the large amplitude plasma waves. An effective collision frequency corresponding to the collective heating is very large; $\nu^* \approx 0.06\omega_0$, where ν^* describes the rate at which the plasma energy increases with time in the nonlinear state.

The computer simulations emphasize another very important feature of the anomalous heating. Figure 9.4 shows a typical heated electron velocity distribution calculated with the particle code. The heating has been principally a production of very high velocity tails on the distribution function [9]. This generation of very high velocity electrons takes place since large amplitude electron plasma waves readily accelerate particles out to their phase velocity.

The physics of the nonlinear saturation can be very rich. There are a number of different regimes depending on the pump field intensity. When the plasma is strongly driven ($eE_0/m\omega_{\mathrm{pe}}v_e \gtrsim 1$), the dominant process is simply strong electron trapping in the most unstable plasma wave. Strong trapping occurs when many electrons are nonlinearly brought into resonance with the wave. A large energy transfer then occurs, as the electrons are efficiently accelerated by the wave.

We can use the calculations of the wave breaking amplitude to estimate the saturation in the simulations, considering an example in the trapping regime: $eE_0/m\omega_0v_e = 1.0$ and $\omega_0 = 1.04\omega_{\mathrm{pe}}$. Linear theory applied to this case predicts that the most unstable plasma wave has a wave number $\simeq 0.25\omega_{\mathrm{pe}}/v_e$ for the electron-ion mass ratio of 0.01 used in this simulation. Equation (9.28) then predicts that strong trapping onsets when $eE/m\omega_{\mathrm{pe}}v_e \simeq 0.8$, which compares reasonably well with the computed value of $eE/m\omega_{\mathrm{pe}}v_e \simeq 0.6$ at saturation.

A simple estimate of the anomalous heating rate can also be given. We estimate the energy transfer from the external driver to the electron plasma oscillations as $2\gamma \langle E_W^2 \rangle/4\pi$, where γ is the linear growth rate and $\langle E_W^2 \rangle/4\pi$ is the energy density of the plasma oscillations. The transfer of

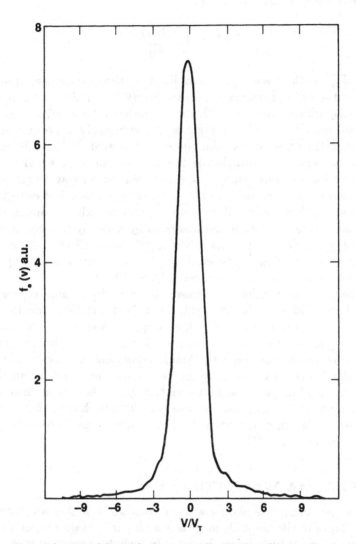

Figure 9.4   A typical heated electron velocity distribution from a particle simulation of a plasma driven by an electric field oscillating near ω_{pe}.

energy to the particles is given by our definition of the anomalous heating rate as $\nu^* E_0^2 / 8\pi$. When the plasma waves saturate, these energy flows balance. Hence, we estimate ν^* as

$$\nu^* = 4\gamma \frac{\langle E_W^2 \rangle}{E_0^2} , \qquad (9.29)$$

where $\langle E_W^2 \rangle$ is the mean square amplitude of the electric field at saturation. For the example discussed above, Eq. (9.29) predicts $\nu^* \simeq 0.04\omega_{pe}$, again comparing reasonably with the computed value of $\nu^* = 0.06\omega_{pe}$.

There are other nonlinear regimes. A particularly important one obtains when the amplitude of the pump field is weaker. Then the excited plasma waves obtain an amplitude $E \sim E_0$ without trapping. Hence they in turn act like efficient "pumps" to drive even shorter wavelength plasma waves, and so on. The net result is a cascade (collapse) of energy from long wavelength waves to short wavelength ones which Landau damp. Again the saturated state is characterized by a steady transfer of energy from the pump field to plasma waves to a heated tail of electrons. This nonlinear transfer of energy to shorter wavelength waves is also important in the evolution of beam-driven instabilities [10–12].

A reduced description was shown to reproduce these moderately-pumped simulations quite well [13]. In this description, the two fluid equations were used to describe the coupled evolution of the electron plasma waves and the ion fluctuations. Simultaneously the electron distribution function (and hence the Landau damping) was evolved by solving a diffusion equation with the diffusion coefficient made a function of the electric field amplitudes. And even though the field structures locally became quite spiky [14], test particle calculations showed that diffusion was a reasonable approximation for the coarse-grained evolution of the distribution function [15].

9.6   PLASMA WAVE COLLAPSE

The tendency of intense plasma waves to cascade to higher wave numbers or to collapse to shorter scale lengths is a very important property of this turbulence. Weak turbulence theory only includes processes such as the ion acoustic decay instability and stimulated scattering on the particles, which down-shift the wave frequency and so transfer energy to longer wavelength (higher phase velocity) waves [16]. In the absence of sufficient collisional damping, wave energy in a driven plasma would indefinitely accumulate at long wavelengths. However, when $E^2/4\pi n\theta_e \gg k^2\lambda_{De}^2$, nonlinear contributions to the wave dispersion relation begin to exceed thermal corrections. Here E is the electric field and k a typical wavenumber of the plasma oscillation, θ_e is the electron temperature, and λ_{De} is the electron Debye length. Weak turbulence theory then no longer applies and wave energy can indeed be coupled into shorter wavelength oscillations, where Landau damping provides an energy sink. The oscillating two stream instability discussed in Chapter 6 is an excellent example of this generation of higher wavenumber plasma waves.

It's very instructive to consider the nonlinear processes in space rather than in a Fourier representation. A local region of intense field expels plasma via the ponderomotive force, forming a density cavity which further localizes and intensifies high frequency oscillations. In two or three dimensions (or in strongly-driven one-dimensional plasmas), the resulting cavity plus its self-consistent high frequency oscillation continues to collapse until efficient damping of the high frequency oscillation onsets. This Landau (or transit-time) damping onsets when the cavity size is of order $10\text{--}20\lambda_{De}$.

The basic theory of electron plasma wave collapse was developed by Zakharov [17]. The analysis is based on a generalization of the coupled equations for the plasma waves and the ion waves discussed in Chapter 6. First we explicitly remove the high frequency time dependence at ω_{pe} i.e., let

$$u_{eh} = \Re \, \bar{u}_{eh}(x,t) \, e^{-i\omega_{pe}t},$$

where u_{eh} is the oscillation velocity of the electron fluid in the high frequency electrostatic field. Equation (6.19) is then readily generalized:

$$\left(i\frac{\partial}{\partial t} + \frac{3}{2}\frac{v_e^2}{\omega_{pe}}\nabla^2 \right)\bar{u}_{eh} = \frac{\omega_{pe}}{2}\frac{n_{e\ell}}{n}\bar{u}_{eh}. \tag{9.30}$$

Note that the driving term on the right hand side of this equation simply represents the coupling of the plasma wave with $n_{e\ell}$, the low frequency fluctuation in the electron density. Likewise, Eq. (6.28) becomes

$$\left(\frac{\partial^2}{\partial t^2} - v_s^2\frac{\partial^2}{\partial x^2} \right)n_{e\ell} = \frac{Z\,m}{4\,M}\,n\frac{\partial^2}{\partial x^2}\bar{u}_{eh}^2, \tag{9.31}$$

where v_s is the ion sound velocity. Equations (9.30) and (9.31) are called the Zakharov equations. The physics of the coupling is clear from our discussion in Chapter 6, where a linearized version of these equations was used to derive the oscillating two stream and ion acoustic decay instabilities.

If ion inertia is neglected, Eq. (9.31) shows that $n_{e\ell}/n = -\bar{u}_{eh}^2/4v_e^2$. Substitution for $n_{e\ell}$ into Eq. (9.30) then gives a nonlinear Schrödinger equation [18]. However, neglect of ion inertia is an extremely restrictive assumption which is quickly violated in the collapse process. The rate (γ_c) at which the high frequency field localizes is proportional to its intensity, which in turn is proportional to the depth of the cavity in this limit. In particular, $\gamma_c/\omega_p \sim \bar{u}_{eh}^2/4v_e^2 \sim n_{e\ell}/n$. Localization of the fields to the cavity requires that the decrease of the plasma wave frequency due to the depression in density be compensated by the increase due to thermal dispersion i.e., $k^2\lambda_{\mathrm{De}}^2 \sim n_{e\ell}/n$. The condition for ignoring ion inertia is $\gamma_c \ll kv_s$. If we substitute the above estimates for γ_c and k, this condition becomes $n_{e\ell}/n \ll Z\,m/M$, where m/M is the electron-ion mass ratio and Z is the ion charge state.

The Zakharov equations admit of solitary wave solutions in one dimension [19–21]. The width of the soliton is related to its amplitude since the nonlinearity is balanced by thermal dispersion. However, such solitons are unstable to two-dimensional perturbations [22,23]. Numerical studies show a collapse to smaller scale lengths, followed by so-called burn-out due to damping by the particles. After burn-out of the high frequency field, the unsupported cavity breaks up into ion acoustic waves which serve as a seed for additional coupling in a driven plasma. A variety of self-similar solutions have been derived to describe the collapse stage, and simulations to isolate the collapse have been carried out [24].

Localized regions of intense high frequency fields within cavities have been observed in a number of experiments in low density laboratory plasmas [25–30]. In some of these experiments [30], the three-dimensional collapse of beam-driven plasma waves has been measured. Although much remains to be understood, the general picture of plasma wave turbulence as a set of randomly occurring collapsing cavities is clearly a very fruitful one. Several reviews of the ongoing work on plasma wave collapse and strong plasma wave turbulence are now available [31–35].

References

1. Stix, T. H., *The Theory of Plasma Waves*. McGraw-Hill, New York, 1962.

2. O'Neil, T. M., Collisionless damping of nonlinear plasma oscillations, *Phys. Fluids* **8**, 2255 (1965).

3. Kruer, W. L. and J. M. Dawson, Sideband instability, *Phys. Fluids* **13**, 2747 (1970).

4. Dawson, J. M., Nonlinear electron oscillations in a cold plasma, *Phys. Rev.* **113**, 383 (1959).

5. Coffey, T. P., Breaking of large amplitude plasma oscillations, *Phys. Fluids* **14**, 1402 (1971).

6. Kruer, W. L., P. K. Kaw, J. M. Dawson and C. Oberman, Anomalous high-frequency resistivity and heating of a plasma, *Phys. Rev. Letters* **24**, 987 (1970).

7. Kruer, W. L. and J. M. Dawson, Anomalous high-frequency resistivity of a plasma, *Phys. Fluids* **15**, 446 (1972).

8. DeGroot, J. S. and J. I. Katz, Anomalous plasma heating induced by a very strong high-frequency electric field, *Phys. Fluids* **16**, 401 (1973).

9. Dreicer, H., R. Ellis and J. Ingraham, Hot electron production and anomalous microwave absorption near the plasma frequency, *Phys. Rev. Letters* **31**, 426 (1973).

10. Thode, L. E. and R. N. Sudan, Two-stream instability heating of plasmas by relativistic electron beams, *Phys. Rev. Letters* **30**, 732 (1973).

11. Kainer, S., J. M. Dawson, and T. Coffey, Alternating current instability produced by the two-stream instability, *Phys. Fluids* **15**, 2419 (1972).

12. Papadopoulous, K., Nonlinear stabilization of beam plasma interactions by parametric effects, *Phys. Fluids* **18**, 1769 (1975).

13. Thomson, J. J., R. J. Faehl and W. L. Kruer, Mode-coupling saturation of the parametric instability and electron heating, *Phys. Rev. Letters* **31**, 918 (1973).

14. Valeo, E. J. and W. L. Kruer, Solitons and resonant absorption, *Phys. Rev. Letters* **33**, 750 (1974).

15. Katz, J. I., J. Weinstock, W. L. Kruer, J. S. DeGroot, and R. J. Faehl, Turbulently heated distribution functions and perturbed orbit theory, *Phys. Fluids* **16**, 1519 (1973).

16. Tsytovich, V. N., *Nonlinear Effects in Plasma*. Plenum Press, New York 1970.

17. Zakharov, V. E., Collapse of Langmuir waves, *Sov. Phys. JETP* **35**, 908 (1972).

18. Morales, G. S., Y. C. Lee and R. B. White, Nonlinear Schrodinger equation model of the oscillating two stream instability, *Phys. Rev. Letters* **32**, 457 (1974).

19. Rudakov, L. I., Deceleration of electrons with a high level of Langmuir turbulence, *Sov. Phys Dokl.* **17**, 1166 (1973).

20. Kingsep, A. S., L. I. Rudakov and R. N. Sudan, Spectra of strong Langmuir turbulence, *Phys. Rev. Letters* **31**, 1482 (1973).

21. Nishikawa, K., H. Hojo and K. Mima, Coupled nonlinear electron-plasma and ion-acoustic waves, *Phys. Rev. Letters* **33**, 148 (1974).

22. Denavit, J., N. R. Pereira and R. N. Sudan, Two-dimensional stability of Langmuir solitons, *Phys. Rev. Letters* **33**, 1435 (1974).

23. Degtyarev, L. M., V. G. Nakhankov and L. I. Rudakov, Dynamics of the formation and interaction of Langmuir solitons and strong turbulence, *Sov. Phys. JETP* **40**, 264 (1975).

24. Anisimov, S. I., M. A. Berezovskii, M. F. Ivanov, I. V. Petrov, A. M. Rubenchik and V. E. Zakharov, Computer simulation of Langmuir collapse, *Phys. Letters* **92A**, 32 (1982).

25. Kim, H. C., R. L. Stenzel and A. Y. Wong, Development of cavitons and trapping of RF fields, *Phys. Rev. Letters* **33**, 886 (1974).

26. Antipov, S. V., M. V. Nezlin, E. N. Snezhkin and A. S. Trubnikov, Excitation of Langmuir solitons by monoenergetic electron beams, *Sov. Phys. JETP* **49**, 797 (1979).

27. Eggleston, D., A. Y. Wong and C. B. Darrow, Development of two-dimensional structure in cavitons, *Phys. Fluids* **25**, 257 (1982).

28. Cheung, P. Y., A. Y. Wong, C. B. Darrow and S. J. Qian, Simultaneous observation of caviton formation, spiky turbulence, and electromagnetic radiation, *Phys. Rev. Letters* **48**, 1348 (1982).

29. Leung, P., M. Q. Tran and A. Y. Wong, Plasma wave collapse generated by the interaction of two oppositely propagating electron beams with a plasma, *Plasma Phys.* **24**, 567 (1982).

30. Wong, A. Y. and P. Y. Cheung, Three dimensional self-collapse of Langmuir waves, *Phys. Rev. Letters* **52**, 1222 (1984).

31. Goldman, M. V., Strong turbulence of plasma waves, *Rev. Mod. Phys.* **56**, 709 (1984).

32. Rubenchik, A. M., R. Z. Sagdeev and V. E. Zakharov, Collapse versus cavitons, *Comm. Plasma Phys. Cont. Fusion* **9**, 183 (1985).

33. Zakharov, V. E., Collapse and self-focusing of Langmuir waves; in *Handbook of Plasma Physics*, Vol. II (A. Galeev and R. N. Sudan, eds.), p.81–122. North Holland, Amsterdam 1984.

34. Shapiro, V. D. and V. I. Shevchenko, Strong turbulence of plasma oscillations, *ibid*, p.123–182.

35. Russell, D., D. F. DuBois, and H. A. Rose, Collapsing caviton turbulence in one dimension, *Phys. Rev. Letters* **56**, 838 (1986).

CHAPTER 10

Density
Profile
Modification

Studies of light-plasma interactions in a plasma with a uniform density are very valuable for understanding many aspects of the physics and have some direct applications to heating of low density, magnetically-confined plasmas. However, laser-produced plasmas are typically rather inhomogeneous. The light-plasma interactions take place in an expanding corona blowing off from an irradiated target. The plasma inhomogeneity affects the mix of the interaction processes. In turn, the interactions can significantly modify the plasma inhomogenity and temperature.

Steepening of the density profile by intense laser light near its critical density is an important example of the interplay between the plasma and the light. As normally incident light reflects at the critical density, twice its pressure is transmitted to the plasma via the ponderomotive force. The plasma expansion is perturbed, leading to a local steepening of the density profile. The profile modification can be enhanced for obliquely-incident, p-polarized light by the ponderomotive force of the resonantly-generated electrostatic waves. As we will see, the profile modification can be substantial and can play an important role in determining both the mix and the scaling of interaction processes near the critical density.

To gain insight into the profile modification, we will first consider a freely expanding plasma and then develop a simple model for profile steepening by normally incident light reflecting at its critical density. Lastly we will briefly examine some simulations of resonance absorption including density profile modification.

10.1   FREELY EXPANDING PLASMA

We begin by deriving a self-similar solution to describe one-dimensional expansion of a planar, isothermal plasma. We again use the two-fluid equations to describe the electrons and ions. With the neglect of electron inertia, the electron momentum equation simply determines the electric field:

$$n_e \, e E \, = \, -\nabla p_e \,, \tag{10.1}$$

where n_e is the electron density and p_e the electron pressure. This electric field transmits the electron pressure to the ions.

The continuity and force equations for the ion fluid give

$$\frac{\partial n}{\partial t} \, + \, \frac{\partial}{\partial x}(nu) \, = \, 0 \tag{10.2}$$

$$\frac{\partial u}{\partial t} \, + \, u \frac{\partial u}{\partial x} \, = \, \frac{Z \, eE}{M} \, - \, \frac{\nabla p_i}{n \, M} \,, \tag{10.3}$$

where n and u are the ion density and flow velocity, Z the ion charge state, M the ion mass and p_i the ion pressure. We next substitute from Eq. (10.1) into Eq. (10.3), neglect the ion pressure relative to the electron pressure, and take $n_e \simeq Zn$, which is an excellent approximation for length scales much greater than the electron Debye length. With an isothermal equation of state for the electrons, Eq. (10.3) becomes

$$\frac{\partial u}{\partial t} \, + \, u \, \frac{\partial u}{\partial x} \, = \, -c_s^2 \frac{1}{n} \frac{\partial n}{\partial x} \,, \tag{10.4}$$

where $c_s = (ZT_e/M)^{1/2}$ is the well-known ion sound velocity and T_e is the electron temperature.

A self-similar solution describing the plasma expansion can be readily found from Eqs. (10.2) and (10.4) by letting $n = f(x/t)$ and $u = g(x/t)$, where f and g are functions to be determined. These equations then give

$$f'\left(g - \frac{x}{t}\right) + f\,g' = 0 \tag{10.5}$$

$$g'\left(g - \frac{x}{t}\right) + c_s^2\,\frac{f'}{f} = 0\,, \tag{10.6}$$

where the prime denotes the derivative with respect to x/t.

Straightforward manipulations yield $g = (x/t) + c_s$ and $f'/f = -c_s^{-1}$. Hence the self-similar solution is

$$u = c_s + \frac{x}{t} \tag{10.7}$$

$$n = n_0 \, \exp\left(-\frac{x}{c_s t}\right), \tag{10.8}$$

where n_0 is the density at $x = 0$. Note that the density gradient length increases with time i.e., $L = n/(\partial n/\partial x) = c_s t$. Note also that, in a frame moving with a point of constant density, $u = c_s$. In other words, the plasma flows through a point of constant density at the sound speed.

10.2 STEEPENING OF THE DENSITY PROFILE

If this expanding plasma is pushed on at a preferred location (for example, at $n = n_{cr}$), the density profile will be locally steepened. The simplest example of this local steepening is that due to the momentum deposition of normally incident light reflecting at its critical density. The basic idea is that twice the pressure of the light wave is taken up by the plasma near the reflection point, and this local momentum deposition steepens the density profile near the critical density [1].

It is instructive to develop a simple model of this profile steepening. In particular, we consider a normally incident light wave reflecting from the critical surface of an isothermal, freely-expanding, collisionless plasma. Again adopting a two-fluid description and assuming planar geometry, we easily obtain equations for the density (n) and flow velocity (u) of the plasma [2–4]. The analysis parallels that discussed in the previous section, with the inclusion of the ponderomotive force exerted by the light wave on the plasma.

If we include the ponderomotive force, Eq. (10.1) becomes

$$n_e \, e \, E \; = \; - \nabla \, p_e \; - \; \frac{n_e m \, \nabla v_w^2}{4} \,, \qquad (10.9)$$

where v_w is the velocity of oscillation of an electron in the electric field of the light wave. Eq. (10.4) now becomes

$$\frac{\partial u}{\partial t} + u \frac{\partial u}{\partial x} = -\frac{c_s^2}{n} \frac{\partial n}{\partial x} - \frac{Z \, m}{4 \, M} \frac{\partial}{\partial x} v_w^2 \,, \qquad (10.10)$$

where c_s is the ion sound velocity, Z the ion charge state, and M the ion mass.

Anticipating that the profile will be steepened from a density $n_1 < n_{\text{cr}}$ to a density $n_2 > n_{\text{cr}}$, we use Eqs. (10.2) and (10.10) to express the variation in density and flow velocity in the frame moving with the steepened surface. In this frame, we have

$$\frac{\partial}{\partial x}(n \, u) = 0 \qquad (10.11)$$

$$\frac{\partial}{\partial x}\left(\frac{u^2}{2}\right) = -c_s^2 \frac{\partial}{\partial x} \ln n - \frac{Z \, m}{4 M} \frac{\partial}{\partial x} v_w^2 \,. \qquad (10.12)$$

If we normalize the flow velocity to the sound speed and substitute from Eq. (10.11) into Eq. (10.12), we obtain

$$(1 - u^2) \frac{1}{n} \frac{\partial n}{\partial x} + \frac{\partial}{\partial x}\left(\frac{v_w^2}{4 v_e^2}\right) = 0 \,, \qquad (10.13)$$

where v_e is the electron thermal velocity. Since the density gradient remains finite, it is clear that the sonic point ($u = 1$) must be at the maximum of the field of the standing light wave i.e., where $\partial v_w / \partial x = 0$. Integrating Eqs. (10.11) and (10.13), we then readily obtain

$$\frac{n_s^2}{n^2} + 2 \ln\left(\frac{n}{n_s}\right) + \frac{v_w^2}{2 v_e^2} = 1 + \frac{v_{\max}^2}{2 v_e^2} \,. \qquad (10.14)$$

Here n_s is the density at the sonic point and $v_{\max}$ is the value of v_w at the maximum of the standing wave. There are two solutions of Eq. (10.14): $n_1 < n_s$ ($u_1 > 1$) which corresponds to a lower density plateau and $n_2 > n_s$ ($u_2 < 1$), which corresponds to the upper density shelf. A schematic of the steepened density profile is shown in Fig. 10.1.

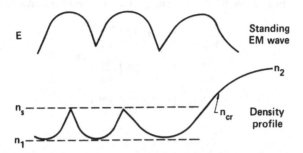

Figure 10.1   A schematic of the ponderomotively-steepened density profile.

To make further progress, we must now relate the density at the sonic point to the critical density (n_{cr}) by considering the solution for the standing electromagnetic wave. A crude treatment will allow us to obtain analytic estimates. We approximate the density profile as locally linear from n_s to n_{cr} with a density scale length of L and express the electric field E by the well-known Airy function solution discussed in Chapter 3:

$$E = \alpha \, \mathrm{A}_i \left[\left(\frac{\omega^2}{c^2 L} \right)^{1/3} (L - x) \right] , \qquad (10.15)$$

where α is a constant determined by fitting to the incoming light wave. This locally linear assumption would be a reasonable approximation whenever $v_{os}/v_e \ll 1$, where v_{os} is the oscillation velocity of an electron in the free space value of the electric field of the light. The assumption fails when $(v_{os}/v_e)^2 \gg 1$, since the jump in density becomes too large. Matching the peak of the Airy function solution to the field at the sonic point, we then obtain $v_w^2(n = n_{cr}) \simeq 0.44 v_{max}^2$. In addition, we note that the sonic point and the critical point are separated by $\Delta x \simeq (c^2 L/\omega^2)^{1/3}$. Equation (10.14) then becomes

$$\frac{n_s^2}{n_{cr}^2} + 2 \ln\left(\frac{n_{cr}}{n_s} \right) = 1 + 0.28 \frac{v_{max}^2}{v_e^2} \qquad (10.16)$$

where $v_{max}^2 \simeq 3.7 v_{os}^2 \left(\omega L/c \right)^{1/3}$ and $n_{cr}/L = (n_{cr} - n_s) \left(\omega^2/c^2 L \right)^{1/3}$.

The solutions for n_1, n_s, and n_2 are now straightforward. (For n_1 and n_2, $v_w = 0$.) For weak fields ($v_{os}/v_e \lesssim 0.1$) analytic results can be given:

$$\frac{n_s}{n_{cr}} \simeq 1 - 0.77 \left(\frac{v_{os}}{v_e}\right)^{0.8}$$

$$\frac{n_s}{n_1} \simeq 1 + 0.97 \left(\frac{v_{os}}{v_e}\right)^{0.8}$$

$$\frac{n_s}{n_2} \simeq 1 - 0.97 \left(\frac{v_{os}}{v_e}\right)^{0.8}$$

$$\frac{\omega L}{c} \simeq 1.5 \left(\frac{v_{os}}{v_e}\right)^{-1.2} . \tag{10.17}$$

Note that the jump in the density scales as a fractional power of the intensity. For more intense fields, numerical solutions of the transcendental equations are required. Results for the steepened scale length as a function of v_{os} are shown in Fig. 10.2. In one interesting regime that is typical of many current applications $(0.1 < (v_{os}/v_e) < 1.)$, $(\omega L/c) \sim 2(v_{os}/v_e)^{-1}$. Numerical solutions for n_1 and n_2 including the detailed structure of the

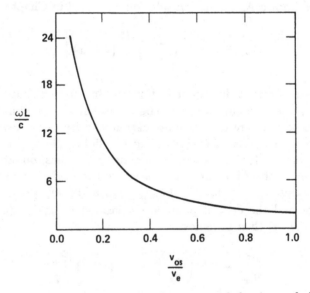

Figure 10.2   Model predictions for the steepened density scale length as a function of v_{os}/v_e. See *Estabrook and Kruer*, (1983).

field are given in Ref. [2].

Since a freely expanding, planar plasma flows through a point of constant density at the sound speed, the momentum deposition at the critical density resonantly perturbs the flow. Hence the profile is steepened over a significant range of densities even for relatively low intensity light. As we will see in the next section, such a profile modification can have a significant effect on the coupling processes near the critical density.

10.3   RESONANCE ABSORPTION WITH DENSITY PROFILE MODIFICATION

Let us conclude our discussion of density profile modification with a more complicated example, which illustrates the nonlinear interplay between resonance absorption of an obliquely incident, p-polarized light wave and profile modification. In this case, the steepening of the density profile is generated both by the pressure of the reflecting, obliquely incident light wave and by the pressure of an intense, resonantly-generated electrostatic field near the critical density [5–10]. We can see the essential features of the nonlinear evolution by examining some computer simulations of resonance absorption.

These simulations [5] are carried out with a two-dimensional code which solves the complete set of Maxwell's equations and includes relativistic particle dynamics. Plane light waves are propagated from vacuum into an inhomogeneous slab of plasma. Variations are followed both along the propagation vector of the light and along its electric vector, which allows for resonance absorption and for the generation of parametric instabilities. Reflected light waves are allowed to freely pass out of the system. Particle boundary conditions are chosen to model a freely expanding plasma adjacent to a reservoir of constant temperature plasma. The initial density varies with x (the direction normal to the slab) from zero to a supercritical value. A region of vacuum is included adjacent to the low density boundary to allow for free expansion of the plasma. Particles impinging on the high density boundary are replaced with equal incoming flux distributed according to $v_x f_m(v)$, where v_x is the component of the velocity normal to the boundary and $f_m(v)$ is the initial Maxwellian velocity distribution. The plasma evolution is followed until a quasi-steady state has been established.

A typical simulation will again illustrate the principal effects. In this example, p-polarized light is incident at an angle of 24° onto an initial density profile which rises linearly from 0 to 1.7 n_{cr} in a distance of $3\lambda_0$ (where λ_0 is the free space wavelengths). The free space amplitude of the electric field of the light is $eE/m\omega_0 c = 0.09$, which corresponds to an intensity of $I\lambda_0^2 \simeq 10^{16}$ W-μ^2/cm^2. The initial electron temperature is 4 keV, and the ion-electron mass ratio is 100.

After the light wave penetrates to its turning point, an electrostatic field is resonantly excited at the critical density. The magnitude of this field initially grows linearly in time, becoming more and more localized to the critical density surface, as expected from the discussion of resonance absorption in Chapter 4. Finally the resonantly-driven field becomes sufficiently intense and localized that electrons can be accelerated through it in one oscillation period, a process called wavebreaking. Physically, wavebreaking corresponds to the onset of strong electron "trapping" in the localized oscillating field. At wavebreaking, electrons which enter the oscillating field with the proper phase are efficiently heated, taking energy from the driven field and saturating its growth.

The feedback of these intense fields (and the concomitant localized heating) on the plasma density profile is a crucial feature of the long-time evolution of the coupling. The pronounced profile modification is demonstrated in Fig. 10.3, which shows three snapshots of the density profile as it evolves from its initial linear profile to a quasi-steady, very steepened

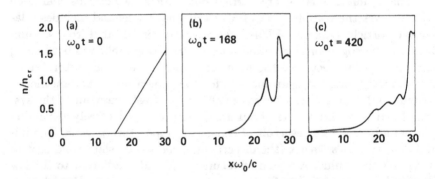

Figure 10.3   The ion density profile at three different times from a simulation of resonance absorption: (a) the initial profile, (b) the profile after the resonantly-driven field has grown, and (c) the asymptotic profile which shows a characteristic step-plateau feature. See *Estabrook et al.*, (1975).

profile. The ponderomotive force due to the intense, localized electrostatic field ejects plasma, digging a hole in the plasma density at the critical surface. The plasma ejected towards the vacuum expands away, leaving a locally steepened density profile which is supported by the pressure of both the localized electrostatic wave and the reflecting light wave.

This profile steepening has important consequences for the mix of absorption processes. In particular, resonance absorption becomes important for a wide range of angles of incidence. This effect is demonstrated in Fig. 10.4, which is a plot of the fractional absorption of p-polarized light (after the profile steepening) versus angle of incidence as computed in a series of simulations with the same initial plasma conditions as the sample simulation. Note that the absorption peaks at about 50% for a sizable angle of incidence ($\theta_{max} \simeq 24°$) and is quite large over a broad range of angles ($\Delta\theta \sim \theta_{max}$). This is qualitatively as expected from our simple theoretical discussion of resonance absorption. In addition, parametric instabilities near the critical density (discussed in the previous

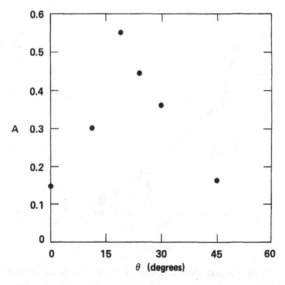

Figure 10.4   The fractional laser light absorption after profile steepening versus angle of incidence as computed in a series of simulations. See *Estabrook et al.*, (1975).

chapter) are strongly limited, since there's a very small region of plasma in which these instabilities can operate. Note that the absorption is only about 15% for normally incident light.

Finally, the profile steepening strongly reduces the heated electron energies due to the resonantly-generated wave. As the wave becomes large enough to nonlinearly interact with the electrons, a small fraction of the electrons (those entering the wave with the proper phases) are strongly heated to an effective temperature of order mv_w^2 where v_w is the oscillation velocity of an electron in the resonantly-driven wave ($v_w = eE/m\omega_0$). As is apparent from our discussion of nonlinear wave-particle interactions, the resonantly-driven field decreases in amplitude as the profile steepens. Physically, the wave then has a smaller spatial extent which corresponds to a lower effective phase velocity. Hence it "traps" electrons at a lower amplitude and heats them to a lower energy.

As expected, the heating via the localized electron plasma oscillation produces a population of suprathermal electrons. Figure 10.5 shows

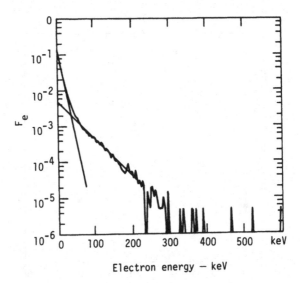

Electron energy – keV

Figure 10.5   The heated electron distribution function from a simulation of resonance absorption.

the electron distribution computed in the steepened, nonlinear state in a sample simulation. The distribution is composed of a relatively cold main body plus a quasi-Maxwellian heated tail.

Self-consistent steepening of the density profile can play an important role in many other laser plasma processes. For example, the two-plasmon-decay instability occurs for a narrow range of densities near one-fourth the critical density. As will be discussed in the next chapter, a local steepening of the profile can help limit this instability. Calculations of collisional absorption must also take profile steepening into account. Even neglecting ponderomotive forces, the density profile is modified by temperature changes driven by the localized heating which occurs on a length scale comparable to the collisional absorption length. This ablative steepening depends on the details of the electron transport.

References

1. Kidder, R. E., Interaction of intense photon beams with plasmas (II);
   in *Proceedings of the Japan-U.S. Seminar on Laser Interaction with Matter*, (C. Yamanaka, ed.). Tokyo International Book, Tokyo, 1973.

2. Lee, K., D. W. Forslund, J. M. Kindel and E. L. Lindman, Theoretical derivation of laser-induced plasma profiles, *Phys. Fluids* **20**, 51 (1977).

3. Takabe, H. and P. Mulser, Self-consistent treatment of resonance absorption in a streaming plasma, *Phys. Fluids* **25**, 2304 (1982).

4. Estabrook, K. and W. L. Kruer, Parametric instabilities near the critical density in steepened density profiles, *Phys. Fluids* **26**, 1888 (1983).

5. Estabrook, K. G., E. J. Valeo and W. L. Kruer, Two-dimensional relativistic simulations of resonant absorption, *Phys. Fluids* **18**, 1151 (1975).

6. Forslund, D., J. Kindel, K. Lee, E. L. Lindman and R. L. Morse, Theory and simulations of resonant absorption in a hot plasma, *Phys. Rev. A* **11**, 679 (1975).

7. DeGroot, J. S. and J. Tull, Heated electron distributions from resonant absorption, *Phys. Fluids* **18**, 672 (1975).

8. Albritton, J. R. and A. B. Langdon, Profile modification and hot electron temperature from resonant absorption at modest intensity, *Phys. Rev. Lett.* **45**, 1794 (1980).

9. Chen, H. H. and C. S. Liu, Soliton generation at resonance and density modification in laser-irradiated plasma, *Phys. Rev. Lett.* **39**, 1147 (1977).

10. Morales, G. J. and Y. C. Lee, Generation of density cavities and localized electric fields in a nonuniform plasma, *Phys. Fluids* **20**, 1135 (1977).

CHAPTER 11

Nonlinear Features of Underdense Plasma Instabilities

In Chapter 10, we discussed some nonlinear phenomena in the neighborhood of the critical density. Let us now examine some nonlinear features of of the light-driven instabilities which can take place in a plasma whose density is significantly below the critical density. In particular, we consider the Brillouin, Raman, two-plasmon decay, and filamentation instabilities. These processes can become significant when large regions of underdense plasma are produced, as is expected for reactor targets in laser fusion. As we will see, both the absorption and the preheat can be strongly affected.

11.1 NONLINEAR FEATURES OF BRILLOUIN SCATTERING

Let us first consider Brillouin scattering. This scattering can most simply be described as the resonant decay of an incident photon into a scattered photon plus an ion sound wave. Hence

$$\omega_0 = \omega_s + \omega_i , \qquad (11.1)$$

where $\omega_0(\omega_s)$ is the frequency of the incident (reflected) light wave and ω_i is the frequency of the ion sound wave. As is apparent from the frequency matching conditions, this process occurs throughout the underdense plasma. In addition, since $\omega_0 \gg \omega_i$, nearly all the energy of an incident photon undergoing this process is transferred to the scattered photon. Hence Brillouin scattering can significantly impact the absorption.

As shown by linear theory, gradients in either expansion velocity or density inhibit the onset of Brillouin scattering. Intensity thresholds due to gradients have been discussed in Chapter 8. These thresholds are very useful for identifying regimes for which Brillouin scatter is not a concern. However, these threshold intensities are often far exceeded, particularly when large regions of underdense plasma are irradiated. The nonlinear behavior [1–9] of this instability then becomes an important issue.

To illustrate nonlinear aspects of the Brillouin instability, let us consider a very simple model problem: the backscattering of a light wave propagating through a slab of underdense plasma with a uniform density. If we postulate that the density fluctuation associated with the ion wave is nonlinearly saturated at some value δn, we can readily calculate the reflectivity [2]. The wave equation which describes the propagation of a light wave with amplitude E through a plasma with density n_e is

$$\left[\frac{\partial^2}{\partial t^2} - c^2 \frac{\partial^2}{\partial x^2} + \omega_{\text{pe}}^2(x) \right] E = 0 , \tag{11.2}$$

where $\omega_{\text{pe}}^2 = 4\pi n_e e^2/m$. We decompose E into an incident and reflected part with slowing varying amplitudes $E_i(x)$ and $E_r(x)$, respectively, and let $n_e = n_0 + \delta n \sin(k_i x - \omega_i t)$. If we substitute n_e into Eq. (11.2) and assume frequency and wave number matching, we obtain

$$\frac{\partial E_r}{\partial x} = -\frac{\omega_{\text{pe}}^2}{4k_0 c^2} \frac{\delta n}{n_0} E_i \tag{11.3}$$

$$\frac{\partial E_i}{\partial x} = -\frac{\omega_{\text{pe}}^2}{4k_0 c^2} \frac{\delta n}{n_0} E_r . \tag{11.4}$$

Since $\omega_i \ll \omega_0$, we have approximated $k_i = 2k_0$, where k_0 is the wave number of the incident light wave.

A conservation law is apparent ($\partial E_i^2/\partial x = \partial E_r^2/\partial x$), which then gives

$$E_i^2 - E_r^2 = E_i^2(0) \, (1 - r) , \tag{11.5}$$

where the reflectivity $r \equiv E_r^2(0)/E_i^2(0)$. Defining $y = E_r(x)/E_i(0)$ and substituting from Eq. (11.5) into Eq. (11.3), we obtain

$$\frac{\partial y}{\partial x} = -\frac{\omega_{pe}^2}{4k_0c^2}\frac{\delta n}{n_0}\sqrt{y^2 + 1 - r}\,. \qquad (11.6)$$

The solution to this standard differential equation is

$$y = \sqrt{1-r}\,\sinh\left[\frac{-\omega_{pe}^2}{4k_0c^2}\frac{\delta n}{n_0}(x - x_0)\right], \qquad (11.7)$$

where x_0 is a constant of integration which is determined by noting that at $x = 0$, $y = \sqrt{r}$. We finally obtain the reflectivity by assuming that $y \simeq 0$ at $x = L$ i.e., $E_r(L) \gg E_i(O)$. Then

$$r = \tanh^2\left(\frac{\omega_{pe}^2 L}{4k_0c^2}\frac{\delta n}{n_0}\right). \qquad (11.8)$$

In order to illustrate the magnitude of the reflectivity, let's estimate the level to which the density fluctuation can be nonlinearly driven. Strong ion trapping (or wave breaking) is one effect commonly invoked to limit the ion wave amplitude. The basic idea has already been discussed in Chapter 9 for electron plasma waves. As the amplitude of the ion wave increases, its potential becomes large enough to nonlinearly bring ions into resonance with the waves. Since such ions are efficiently accelerated by the wave, a strong damping results, which serves to restrict the ion wave amplitude from further increase. If the ions are cold, the trapping condition is simply $Ze\phi = Mv_p^2/2$, where ϕ is the potential, M the ion mass, and v_p the phase velocity of the wave. Neglecting Debye length corrections, the trapping condition corresponds to $\delta n/n_0 \simeq e\phi/\theta_e \simeq 1/2$, which is a large amplitude.

It is important to realize that even a small ion temperature significantly reduces the trapping amplitude. This temperature effect is readily estimated if one assumes a so-called waterbag velocity distribution for the ions. In one-dimension, such a distribution is constant with velocity between $\pm\sqrt{3}v_i$ (v_i is the ion thermal velocity) and zero elsewhere. Since the majority of the ions in a Maxwellian distribution have velocities $< 2v_i$, the waterbag distribution gives a reasonable first approximation for the onset of strong trapping. Trapping now occurs when the fastest ion is nonlinearly brought into resonance with the wave i.e.,

$$Z e \phi = \frac{M}{2} \left(v_p - \sqrt{3} \, v_i \right)^2$$

$$\frac{\delta n}{n_0} \simeq \frac{1}{2} \left(\sqrt{1 + \frac{3\theta_i}{Z\theta_e}} - \sqrt{\frac{3\theta_i}{Z\theta_e}} \right)^2 . \tag{11.9}$$

Here $\theta_i(\theta_e)$ is the ion (electron) temperature, Z is the ion charge, and the Debye length correction to the phase velocity has been neglected. For $\theta_i/Z\theta_e = 0.2$, Eq. (11.9) predicts a fluctuation amplitude of $\delta n/n_0 \simeq 0.12$. Clearly the ion temperature serves to significantly reduce the amplitude, but note that the trapping amplitude is still of order 10%, unless the ions are quite hot (i.e., $\theta_i/Z\theta_e \sim \mathcal{O}(1)$). Strong trapping does not, in general, limit the fluctuation amplitude to a small value.

Using the ion trapping estimate for δn, we can now calculate a reflectivity. As an example, consider a $30\lambda_0$ slab of plasma with a uniform density of $n_0 = 0.33n_{cr}$ and an ion-electron temperature ratio of $\theta_e/Z\theta_i = 0.2$. Here λ_0 is the free-space wavelength of the light. Substituting from Eq. (11.9) into Eq. (11.8), we obtain $r \simeq 94\%$. Even a modest fluctuation amplitude can lead to a sizable reflectivity in a large underdense plasma.

Wavebreaking (or strong trapping) arguments only give an estimate of the amplitude at which a strong damping onsets due to wave-particle interactions. When a significant number of ions are accelerated and sizeable tails develop on the ion distribution function, a more complex description of the nonlinear wave particle interaction is needed. To gain insight into these effects, let's now consider some computer simulations of Brillouin scattering.

Figure 11.1 shows the temporal evolution of the Brillouin back reflection and the energy in the ions computed with a one-dimensional code [3] which treated the ions as particles and the electrons as a fluid. In this example, light with an intensity of $I\lambda_0^2 = 3 \times 10^{15}$ W-μ^2/cm^2 is incident onto a $30\lambda_0$ slab of plasma with an initially uniform density of $n_0 = 0.33n_{cr}$, an electron temperature of 3 keV, and an ion-electron temperature ratio of 0.2. In the simulation, ions reaching the right plasma boundary were re-emitted with the initial thermal temperature, modeling transport of heated ions to a higher density plasma. As shown in Fig. 11.1(a), the reflectivity rapidly proceeds to a large level of about 65%, as the unstable

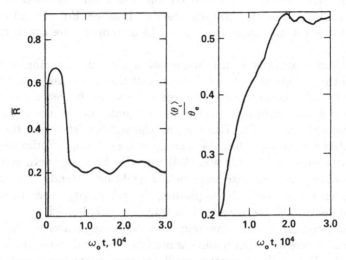

Figure 11.1   Evolution of the reflectivity (time-averaged over many cycles) and the mean ion energy from a computer simulation of Brillouin backscatter.

ion wave grows to a large amplitude and traps ions. However, concomitant with this large reflectivity is a substantial heating of the ions, as shown in Fig. 11.1(b). In detail, this heating consists of the formation of a sizable tail of energetic ions as expected from our discussion of collisionless wave particle heating in Chapter 9. This strong self-consistent distortion of the ion velocity distribution in turn enhances the damping of the ion wave and lowers its amplitude. Finally a quasi-steady state is reached in which the heating of ions by the wave is balanced by their transport out of the underdense plasma. The reflectivity drops to a more modest value of about 25%.

This ion heating or energetic tail formation is an intrinsic feature of the Brillouin scattering. In the scattering process a fraction ω_i/ω_0 of the reflected light energy is deposited into the ion wave and then into the heated ions when the wave damps. We first estimate an effective temperature of the heated ions as $M v_s^2/2$ which is the energy of an ion moving with the phase velocity (v_s) of the wave. The density n_h of the ion tail is then estimated by balancing the energy flux into the ion wave with that carried away by the heated ion tail; i.e.,

$$\frac{n_h \, M \, v_s^3}{2} \approx r \, I \, \frac{\omega_i}{\omega_0} \, . \tag{11.10}$$

Here the flux carried by the heated ions has been described in a free-streaming limit. For the example discussed above, Eq. (11.10) predicts $(n_h/n_0) \approx 0.4$ in the nonlinear state, which compares well with the simulation.

The ion heating is a clear manifestation of the damping of the ion wave in the nonlinear state. If we represent the scattering as a reflection from a heavily damped wave, the reflectivity can be readily calculated in terms of this damping. A heuristic estimate for the damping in the nonlinear state is Landau damping on the self-consistent ion tail (or on the heated main body when the heating is very strong). Estimates of the reflectivity obtained in this way [2,4] compare favorably with simulation results, as well as illustrate some of the qualitative trends. For example, for a given size of underdense plasma, the reflectivity tends to saturate with intensity. The increase in light intensity is balanced by an increase in the self-consistent damping associated with the greater ion heating.

Finally there are other nonlinear mechanisms for limiting the Brillouin instability. Harmonic generation [4–6], quasi-resonant decay of ion waves [7], nonlinear frequency shifts [8,9], and profile steepening can play a significant role in some regimes. Quantitative calculations [10,11] of Brillouin scattering also require consideration of the noise sources as well as of the detailed profile of the density and expansion velocity of the plasma. A significant complication is that partial reflection of the light wave from the critical surface can serve as a noise source in an expanding plasma [12,13]. The angular distribution of the scattering and the competition with other underdense plasma processes such as inverse bremsstrahlung and filamentation are other important issues.

11.2 NONLINEAR FEATURES OF RAMAN SCATTERING

To complement this brief discussion of Brillouin scattering, let us now consider some nonlinear aspects [1,14–20] of the Raman instability. The Raman instability can be most simply characterized as the resonant decay of an incident photon into a scattered photon plus an electron plasma wave. The frequency and wave number matching conditions then are

$$\omega_0 = \omega_s + \omega_{pe}, \qquad \mathbf{k}_0 = \mathbf{k}_s + \mathbf{k}_p, \qquad (11.11)$$

where ω_0 (ω_s) and $\mathbf{k}_0$ ($\mathbf{k}_s$) are the frequency and wave number of the incident (scattered) light wave, and ω_{pe} ($\mathbf{k}_p$) is the frequency (wave number) of the electron plasma wave. Since the minimum frequency of a light wave in a plasma is ω_{pe}, the electron plasma frequency, it is clear that this process requires that $\omega_0 \gtrsim 2\omega_{pe}$ i.e., $n \lesssim n_{cr}/4$, where n is the plasma density and n_{cr} is the critical density. In this process, part of the incident energy is scattered, and part is deposited in the electron plasma wave. This latter portion of the energy in turn will heat the plasma as the electron plasma wave damps. Since the plasma frequency is much greater than the ion acoustic frequency, the Raman instability is clearly not as efficient in scattering the incident laser light as is the Brillouin instability. However, the electron plasma wave which is generated can have a high phase velocity (of order c) and so can produce very energetic electrons when it damps. Since such electrons can preheat the fuel in laser fusion applications, the Raman instability is a particularly significant concern.

As discussed in Chapter 7, the intensity threshold due to a density gradient is rather high but can clearly be exceeded in a large underdense plasma irradiated with intense light. Again, we will use simulations to give us some estimates of what to expect in the nonlinear regime. To most simply explore the nonlinear effects, consider a 1-$\frac{1}{2}$ dimensional particle simulation [14] in which laser light with intensity $I\lambda_0^2 = 2.5 \times 10^{15}$ W-μ^2/cm^2 is propagated through a $127\lambda_0$ region of plasma with a uniform density of $0.1n_{cr}$, an electron temperature of 1 keV, and an electron-ion temperature ratio of three. In this simulation, the back reflection due to the Raman instability builds up to about 15%, accompanied by strong tail heating of the electrons by the electron plasma wave associated with the scatter. (There is also a modest back reflection of $\sim$ 20% due to the Brillouin instability in this example.) The resulting heated electron distribution is shown in Fig. 11.2. Note the heated tail, which is roughly Maxwellian in shape with a characteristic temperature $\theta_H \simeq 13$ keV. Such energetic tail formation is characteristic of heating via a large amplitude electron plasma wave. A useful rule of thumb estimate for the heated temperature found in these strongly-driven simulations is $\theta_H \approx mv_p^2/2$, which is simply the energy of an electron accelerated to the phase velocity v_p of the plasma wave. As can be seen from the frequency and wave number matching conditions, such a temperature depends on both the density and background electron temperature and can easily be of order 50–100 keV, even for backscatter.

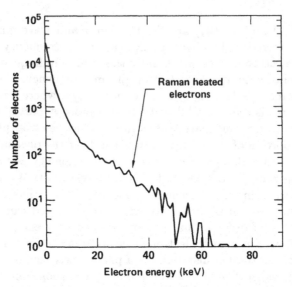

Figure 11.2   The heated electron distribution from a computer simulation of Raman backscatter. See *Estabrook et al.*, (1980).

This self-consistent electron heating is a significant feature of the nonlinear evolution and can play a role in restricting the scatter. It is instructive to estimate the size of the heated tail in the nonlinear state by balancing the energy flux deposited in the electron plasma wave (and hence into heated electrons when the wave damps) with the energy flux carried away by the heated tail. Neglecting background thermal effects and using a free-streaming estimate of the hot electron transport gives

$$r\,I\,\frac{\omega_{\text{pe}}}{\omega_0 - \omega_{\text{pe}}} \approx 0.6\,n_h\,\theta_H \sqrt{\frac{\theta_H}{m}}\,, \tag{11.12}$$

where r is the reflectivity and n_h is the tail density. If we consider backscatter, assume $\omega_{\text{pe}}/\omega_0 \gg 1/2$, and use our estimate of θ_H, we obtain

$$\frac{n_h}{n} \sim 19\,r\left(\frac{v_{\text{os}}}{c}\right)^2\left(\frac{n_{\text{cr}}}{n}\right)^2, \tag{11.13}$$

where n is the plasma density, v_{os} is the oscillation velocity of the electrons in the light wave, and c is the velocity of light. Clearly even a modest reflectivity is sufficient to drive the instability into a regime in which a significant fraction of the electrons are resonant with the plasma wave.

A ball-park estimate of the Raman back reflection in the sample simulation can be obtained by estimating the damping of the plasma wave associated with the hot electron generation. If the damping is crudely modeled as Landau damping on the heated tail, the reflectivity due to this damped plasma wave is readily calculated, giving $r \simeq 10\%$ for the above example. This back-of-the-envelope model illustrates some important features of the nonlinear evolution: the self-consistent generation of hot electrons and their feedback on the instability.

Ion fluctuations can also play an important role [16–21] in Raman scattering. The ion fluctuations are produced either by the Brillouin instability or by collapse of the Raman-generated plasma wave. As discussed in Chapters 6 and 9, an ion fluctuation efficiently couples plasma waves provided $\delta n/n > \Delta\omega/\omega_{pe}$, where δn is the amplitude of the density fluctuation and $\Delta\omega$ is the frequency mismatch between the plasma waves. This energy transfer from the Raman-driven plasma wave into shorter wavelength plasma waves both reduces the level of the primary wave and produces less energetic heated tails. Frequency shifts in the primary plasma wave due to the ion waves may also be significant.

Even when the Raman instability is not operative, an incident light wave can still undergo stimulated scattering on the electrons [22,23]. There also are important multi-dimensional effects, including Raman sidescattering and filamentation of the incident light wave. Two-dimensional simulations [24] using a very intense beam of light emphasize the importance of Raman sidescattering and even show filamentation due to relativistic effects. The latter is accentuated by self-generated magnetic fields driven by the Weibel instability of the heated electrons.

11.3 NONLINEAR FEATURES OF THE TWO-PLASMON DECAY AND FILAMENTATION INSTABILITIES

Let's now consider another instability involving electron plasma waves: the two-plasmon decay or $2\omega_{pe}$ instability. As discussed in Chapter 7, this instability represents the resonant decay of a light wave into two electron plasma waves. The frequency matching condition clearly requires that $\omega_0 \simeq 2\omega_{pe}$ i.e., $n \simeq n_{cr}/4$. The feedback mechanism leading to instability is similar to that already discussed for the Raman instability, except now both growing waves are electron plasma waves. The maximum growth rate in a uniform plasma is the same as that for the Raman instability at $n_{cr}/4$, but now a broad spectrum of plasma waves is unstable.

Because the $2\omega_{pe}$ instability is confined to a narrow range of densities near $n_{cr}/4$, a local nonlinear steepening of the density profile can play an especially important role in the nonlinear evolution. A density profile from an illustrative simulation [25,26] of the $2\omega_{pe}$ instability is shown in Fig. 11.3. In the two-dimensional simulation, laser light with an intensity of $I\lambda_0^2 = 10^{16}$ W-μ^2/cm^2 is incident onto an initially inhomogeneous plasma slab. The initial electron temperature is 1 keV, and the electron-ion mass ratio is 0.01. Note the pronounced steepening which takes place near $n_{cr}/4$ due to the instability-generated plasma waves.

In the simulation, the instability occurs in bursts, as the density profile steepens and relaxes. The averaged absorption in the steepened profile is modest (of order 10%). During periods of instability generation, hot electron tails are formed with an effective temperature of about 100 keV for this strongly-driven example. Ion fluctuations driven by beating of the unstable plasma waves are observed to play an important role in the non-

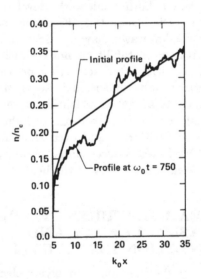

Figure 11.3   A density profile from simulation of the $2\omega_{pe}$ instability. *Langdon et al.* (1979).

linear evolution. The importance of ion density fluctuations is also emphasized by the nonlinear theory [27–29].

Finally we conclude with a brief discussion of the filamentation instability, which can have an important effect on the mix of coupling processes. As discussed in Chapter 6, this instability represents the development of filamentary structure in the intensity profile of a light wave. The instability occurs throughout the underdense plasma and is related to whole-beam self-focusing. Both filamentation and whole-beam self-focusing can be driven by ponderomotive, thermal, or even relativistic effects [30]. The processes can be accentuated by resonantly-enhanced fields [31]. For simplicity, we will here concentrate on ponderomotive filamentation.

To illustrate the rich possibilities introduced by filamentation, consider a two-dimensional simulation [32] in which the temporal evolution of an intense light wave is followed in a doubly-periodic plasma. The background plasma density is $0.31n_{cr}$, the electron temperature is 4 keV, and the intensity of the light wave is $I\lambda_0^2 \simeq 2.3 \times 10^{16}$ W-μ^2/cm^2. The electron-ion mass ratio is 0.01, and the ion temperature is large in order to suppress the competing effects of the Brillouin instability. A small sinusoidal density modulation perpendicular to the direction of the wave propagation and along the direction of the electric field of the light wave serves as an initial perturbation for the growth. This density perturbation and the corresponding modulation in the intensity of the light wave grows in time. When the density in the channel has been depressed to about 0.25 n_{cr}, the laser light decays into intense electrostatic fields, which in turn heat the electrons. A contour plot of the electrostatic potential in the simulation at this time is shown in Fig. 11.4. Note that the electrostatic fields are concentrated in the channel. In other simulations, another type of decay analogous to stimulated Raman scattering was observed.

The competition of filamentation with other processes is a very rich topic. In some two-dimensional simulations [33], it has been found that intense Brillouin sidescattering can suppress filamentation. In other calculations [34], self-focusing of a light wave was arrested by intense Brillouin backscattering which onset as the intensity of the light wave increased. In addition, calculations have shown that filaments can be unstable [35] to bending along the direction of their propagation. A general picture of the role of filaments in laser plasma interactions has not yet emerged.

In summary, in this chapter we have illustrated some important effects produced by intense laser light in a plasma with a density below the

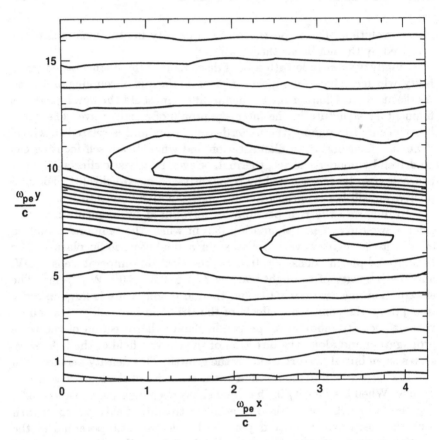

Figure 11.4   Contour plot of the electrostatic potential in the simulation when channel density has been depressed to about $n_{cr}/4$. From *Langdon and Lasinski* (1975).

critical density. Our discussion emphasizes that crucial features of the coupling can depend on the size of the underdense plasma in a laser-irradiated target. The possible consequences of sizable regions of underdense plasma include significant degradation of the absorption and/or the generation of very energetic electrons, which can complicate fusion target design.

References

1. Forslund, D. W., J. M. Kindel, and E. L. Lindman, Plasma simulation studies of stimulated scattering processes in laser-irradiated plasmas, *Phys. Fluids* **18**, 1002 (1975).

2. Kruer, W. L., Nonlinear estimates of Brillouin scatter in plasma, *Phys. Fluids* **23**, 1273 (1980).

3. Kruer, W. L., E. J. Valeo, and K. G. Estabrook, Limitation of Brillouin scattering in plasmas, *Phys. Rev. Lett.* **35**, 1076 (1975).

4. Kruer, W. L. and K. G. Estabrook, Nonlinear behavior of stimulated scatter in large underdense plasmas; in *Laser Interaction and Related Plasma Phenomena*, Vol. 5, p.783–800 (H. Schwarz, H. Hora, M. Lubin, and B. Yaakobi, eds.). Plenum Press, New York, 1981.

5. Silin, V. P. and V. T. Tikhonchuk, Nonlinear saturation of SMBS in a rarefied nonisothermal plasma, *JETP Letters* **34**, 365 (1981).

6. Heikkinen, J. A., S. J. Kartunnen, and R. R. E. Saloman, Ion acoustic non-linearities in stimulated Brillouin scattering, *Phys. Fluids* **27**, 707 (1984).

7. Kartunnen, S. J., J. N. McMullin, and A. A. Offenberger, Saturation of stimulated Brillouin scattering by ion wave decay in a dissipative plasma, *Phys. Fluids* **24**, 447 (1981).

8. Casanova, M., G. Laval, R. Pellat, and D. Pesme, Self-generated loss of coherency in Brillouin scattering and reduction of reflectivity, *Phys. Rev. Lett.* **54**, 2230 (1985).

9. Ikezi, H., K. Schwarzeneggar, A. L. Simons, Y. Ohsawa, and T. Kamimura, Nonlinear self-modulation of ion-acoustic waves, *Phys. Fluids* **21**, 239 (1978).

10. Ramani, A. and C. E. Max, Stimulated Brillouin scattering in an inhomogeneous plasma with broad bandwidth thermal noise, *Phys. Fluids* **26**, 1079 (1983).

11. Colombant, D. G. and W. M. Manheimer, A model of anomalous absorption, backscatter and flux limitation in laser-produced plasmas, *Phys. Fluids* **13**, 2512 (1980).

12. Randall, C. J., J. R. Albritton, and J. J. Thomson, Theory and simulation of stimulated Brillouin scatter excited by nonabsorbed light in laser fusion systems, *Phys. Fluids* **24**, 1474 (1981).

13. Randall, C. J. and J. R. Albritton, Chaotic nonlinear stimulated Brillouin scattering, *Phys. Rev. Lett.* **52**, 1887 (1984).

14. Estabrook, K. G., W. L. Kruer, and B. F. Lasinski, Heating by Raman backscatter and forward scatter, *Phys. Rev. Letters* **45**, 1399 (1980).

15. Biskamp, D. and H. Welter, Stimulated Raman scattering from plasmas irradiated by normally and obliquely incident laser light, *Phys. Rev. Lett.* **34**, 312 (1975).

16. Bonnaud, G., Ion mobility influence on stimulated Raman scattering in homogeneous laser-irradiated plasma, *Laser and Particle Beams* **5**, 101 (1987).

17. Estabrook, K. and W. L. Kruer, Theory and simulation of one-dimensional Raman backward and forward scattering, *Phys. Fluids* **26**, 1892 (1983).

18. Aldrich, C. H., B. Bezzerides, D. F. DuBois, and H. A. Rose, Langmuir nucleation and collapse in stimulated laser light scatter, *Comm. Plasma Phys.* **10**, 1 (1986).

19. Rozmus, W., R. P. Sharma, J. C. Samson, and W. Tighe, Nonlinear evolution of stimulated Raman scattering in homogeneous plasmas, *Phys. Fluids* (1987).

20. Barr, H. C. and G. A. Gardner, Harmonic emission from the quarter critical density surface of laser-produced plasmas; in *Proceedings of the International Conference on Plasma Physics* Vol II, p.265 (Q. Tran and R. J. Verbeek, eds.). École Polytechnique de Lausanne, Lausanne, 1984.

21. Barr, H. C. and F. F. Chen, Raman scattering in a nearly resonant density ripple, *Phys. Fluids* **30**, 1180 (1987).

22. Ott, E., W. M. Manheimer, and H. H. Klein, Stimulated Compton scattering and self-focusing in the outer regions of a laser fusion plasma, *Phys. Fluids* **17**, 1757 (1974).

23. Lin, A. T. and J. M. Dawson, Stimulated Compton scattering of electromagnetic waves in plasma, *Phys. Fluids* **18**, 201 (1975).

24. Forslund, D. W., J. M. Kindel, W. B. Mori, C. Joshi, and J. M. Dawson, Two-dimensional simulations of single-frequency and beat-wave laser-plasma heating, *Phys. Rev. Lett.* **54**, 558 (1985).

25. Langdon, A. B., B. F. Lasinski and W. L. Kruer, Nonlinear saturation and recurrence of the two-plasmon decay instability, *Phys. Rev. Letters* **43**, 133 (1979).

26. Langdon, A. B., B. F. Lasinski, and W. L. Kruer, **ZOHAR** simulations of two-plasmon-decay; in *Lawrence Livermore National Laboratory UCRL-50021-85*, p.2–43 (1986).

27. Chen, H. H. and C. S. Liu, Soliton formation and saturation of decay instability of an electromagnetic wave into two plasma waves, *Phys. Rev. Lett.* **39**, 881 (1977).

28. Kartunnen, S. J., Saturation of parametric instabilities by the nonlinear decay of electrostatic daughter wave, *Plasma Phys.* **22**, 151 (1980).

29. Shapiro, V. D. and V. I. Shevchenko, Strong turbulence of plasma oscillations; in *Handbook of Plasma Physics*, Vol II, p.123–182 (A. A. Galeev and R. N. Sudan, eds.). North Holland, Amsterdam, 1984.

30. Max, C. E., Physics of the coronal plasma in laser fusion fusion targets; in *Laser-Plasma Interaction*, (R. Balian and J. C. Adam, eds.). North Holland, Amsterdam, 1982.

31. Joshi, C., C. E. Clayton, and F. F. Chen, Resonant self-focusing of laser light in a plasma, *Phys. Rev. Lett.* **48**, 874 (1982).

32. Langdon, A. B. and B. F. Lasinski, Filamentation and subsequent decay of laser light in plasmas, *Phys. Rev. Lett.* **34**, 934 (1975).

33. Estabrook, K. G., Critical surface bubbles and corrugations and their implications to laser fusion, *Phys. Fluids* **19**, 1733 (1976).

34. Randall, C. J., Simultaneous self-focusing and Brillouin backscattering of Gaussian laser beams; in *Lawrence Livermore National Laboratory UCRL -50021-79*, p.3–45 (1980).

35. Valeo, E. J., Stability of filamentary structures, *Phys. Fluids* **17**, 1391 (1974).

CHAPTER 12

Electron Energy Transport

In the previous chapters, we have discussed a number of different processes by which laser light heats the plasma surrounding a laser-irradiated target. As we have seen, the energy is deposited primarily into electrons. The rate at which the electrons in turn transport this energy to the higher density colder plasma determines both the efficiency of the implosion and the plasma conditions in the region of deposition. Not surprisingly, the transport of large fluxes of energy in inhomogeneous plasmas is itself a rich and complex topic. We will begin with the classical diffusive calculation of electron heat conduction and then discuss its extension to multigroup, flux-limited diffusion. Lastly, we will indicate some of the additional complications due to self-generated magnetic fields or ion acoustic turbulence and conclude with a brief summary of what experiments have indicated.

12.1 ELECTRON THERMAL CONDUCTIVITY

It's instructive to begin with the classical Spitzer-Harm calculation [1,2] of electron thermal conductivity in a plasma with no magnetic fields. Neglecting hydrodynamic motion and density gradients, we start with the kinetic equation introduced in Chapter 5:

$$\frac{\partial f}{\partial t} + \mathbf{v} \cdot \frac{\partial f}{\partial \mathbf{x}} - \frac{e\mathbf{E}}{m} \cdot \frac{\partial f}{\partial \mathbf{v}} = A \frac{\partial}{\partial \mathbf{v}} \cdot \left[\frac{v^2 \underline{\underline{\mathbf{I}}} - \mathbf{v}\mathbf{v}}{v^3} \cdot \frac{\partial f}{\partial \mathbf{v}} \right] + C_{ee}(f) \,. \quad (12.1)$$

In this expression, f is the electron velocity distribution function, $A = (2\pi n Z e^4/m^2)\ln\Lambda$, and $C_{ee}(f)$ denotes a similar but more complex operator giving electron-electron collisions. Further, n is the plasma density, Z the ion charge-state, and $\ln\Lambda$ is the Coulomb logarithm. In order to carry heat, the distribution function must be warped in the direction of the heat flow. Adopting spherical coordinates, we represent the distribution function as the first two terms of an expansion in Legendre polynomials:

$$f = f_0(v) + f_1(v)\cos\theta \,, \quad (12.2)$$

where $v = |\mathbf{v}|$ and θ is the angle between $\mathbf{v}$ and the direction of the heat flow (also the direction of $\mathbf{E}$). Substituting for $f(v)$ into Eq. (12.1) and collecting the terms proportional to $\cos\theta$, we obtain an equation for $f_1(v)$:

$$\frac{\partial f_1}{\partial t} + v\frac{\partial f_0}{\partial z} - \frac{e}{m}E\frac{\partial f_0}{\partial v} = -\frac{2A}{v^3}f_1 \,. \quad (12.3)$$

Here we have assumed for simplicity a high Z plasma and neglected the effect on $f_1(v)$ of electron-electron collisions relative to electron-ion ones. The steady-state solution is

$$f_1 = -\frac{v^4}{2A}\left(\frac{\partial f_0}{\partial z} - \frac{eE}{mv}\frac{\partial f_0}{\partial v}\right) \,. \quad (12.4)$$

The electric field is determined by the condition that charge neutrality be preserved i.e., $J_z = -e\int \mu f_1 \, \mu v \, d\mathbf{v} = 0$, where $\mu = \cos\theta$. This condition gives

$$\int_0^\infty dv \, v^7 \left(\frac{eE}{mv}\frac{\partial f_0}{\partial v} - \frac{\partial f_0}{\partial z}\right) = 0 \,. \quad (12.5)$$

We next assume that electron-electron collisions maintain the zero-order distribution $f_0(v)$ as a Maxwellian with a local temperature $\theta_e(z)$:

$$f_0(v) = \frac{n}{(2\pi)^{3/2}\, v_e^3(z)} \exp\left[-\frac{m\,v^2}{2\theta_e(z)}\right] , \qquad (12.6)$$

where $v_e^2(z) = \theta_e(z)/m$. Equation (12.5) is now readily integrated to give

$$e\,E = -\frac{5}{2}\frac{\partial\,\theta_e(z)}{\partial z} . \qquad (12.7)$$

Hence f_1 becomes

$$f_1 = f_0 \frac{v^4}{4\,m\,A}\left[\frac{8}{v_e^2(z)} - \frac{v^2}{v_e^4(z)}\right]\frac{\partial\theta_e}{\partial z} . \qquad (12.8)$$

The heat flow Q is obtained by using $f_1(v)$ to evaluate

$$Q = \int \frac{m\,v^2}{2}\,\mu v\,\mu\,f_1\,\mathrm{d}^3\mathbf{v} . \qquad (12.9)$$

After straight-forward integration, we obtain,

$$Q = -\kappa\frac{\partial\theta_e}{\partial z} , \qquad (12.10)$$

$$\kappa = \frac{4\,\theta_e^{5/2}}{Z e^4\, m^{1/2}\,\ln\Lambda} . \qquad (12.11)$$

Note that the thermal conductivity κ is independent of density and is proportional to $\theta_e^{5/2}$. A convenient representation is $\kappa \simeq 14\,n v_e^2/\nu_{ei}$, where ν_{ei} is the collision frequency which describes collisional damping of a light wave as discussed in Chapter 5.

For small to moderate values of Z, it becomes important to directly include electron-electron collisions which, of course, reduce the conductivity. A simple approximation to the numerical results of Spitzer-Harm is given by multiplying κ given in Eq. (12.11) by $g(Z) \simeq (1 + 3.3/Z)^{-1}$.

12.2 MULTIGROUP FLUX-LIMITED DIFFUSION

Although an instructive point of departure, this classical calculation of diffusive heat flow needs to be extended in several important ways. First, since the mean free-path is energy-dependent, a conductivity which averages over a distribution of disparate velocities is clearly inadequate to properly treat the transport of energy into a target. Hence the heat transport is usually modeled as multigroup diffusion [3]. In this description, the electrons are divided into energy groups, with the lowest energy group described as a thermal one, with a Maxwellian velocity distribution. Each group is assigned a diffusion coefficient, and the groups are coupled to one another by self-consistent electric fields imposed by charge neutrality. Such a description is especially important when high energy tails are produced by collective absorption processes.

In general, a second extension is also needed, since the diffusive calculation of the heat flow fails for strong temperature gradients. For example, when $(v_e/\nu_{ei}) \gtrsim 0.1 L_T$ $(L_T^{-1} = \partial \ln \theta_e/\partial z)$, Eq. (12.10) gives $Q \gtrsim n\theta_e v_e$. This result is clearly unphysical, since electrons cannot carry an energy flux greater than their energy density times some typical velocity. In fact, the Spitzer-Harm calculation is expected to fail [4,5] for even smaller heat fluxes (i.e., for longer L_T). We first note from Eqs. (12.8) and (12.9) that electrons with velocities up to about 4–5 v_e contribute significantly to the heat flow. Substituting Eq. (12.10) into Eq. (12.8) gives

$$f_1 = f_0 \frac{Q}{32\,\pi\,n\,\theta_e\,v_e} \left(\frac{v}{v_e}\right)^4 \left[\left(\frac{v}{v_e}\right)^2 - 8\right].$$

Demanding that $|f_1| < f_0$ for $v \sim 4v_e$ then requires that $Q \lesssim nQ_e v_e/5$.

From a physical standpoint, the breakdown of Eq. (12.10) for strong temperature gradients represents the transition to collisionless behavior in which electrons simply free-stream rather than diffuse. Heuristic attempts to match onto this collisionless regime have been made by simply limiting the heat flow to a maximum flux. To illustrate, we again return to the simple example in which electrons are characterized by a temperature θ_e. Then $Q = \min[\kappa\,\partial\theta_e/\partial z,\ f\,n\theta_e v_e]$, where f is the so-called flux limit. Initially f was chosen to be about 0.6, since the maximum energy flux (neglecting fields) carried by electrons with a Maxwellian distribution of velocities is about $0.6\,n\theta_e v_e$. However, clearly the flux limit is just a crude but efficient attempt to describe the heat flow as the classical theory fails.

To properly describe the heat flux in this limit, both numerical calculations of the full kinetic equation (including electron-electron collisions) and Monte-Carlo calculations have been carried out [6–15]. In comparison to the Spitzer-Harm treatment, more angular structure and/or modifications to the zero-order distribution functions have been included. Indeed, the modification of $f_0(v)$ is obviously essential to include in many applications, since classical absorption can generate super-Gaussian velocity distributions as discussed in Chapter 5. Furthermore, the transport itself can strongly modify the zero-order distribution function. These calculations have indicated that the heat flux tends to saturate at a value of order $0.1\, n\theta_e v_e$ but also emphasize that a single flux limit is in general too simple a parameterization. A significant challenge is to now incorporate these insights into improved transport models which are sufficiently economical for routine use in design codes.

12.3   OTHER INFLUENCES ON ELECTRON HEAT TRANSPORT

Thus far we have considered the electron heat transport as determined by Coulomb collisions in a plasma with no magnetic fields. The heat transport becomes an even more complex calculation when one includes self-generated magnetic fields. There are many source terms [16–18] for such fields; the best known is that which arises when the density and temperature gradients are not parallel. This effect can be simply illustrated. If we treat the electrons as a fluid and neglect their inertia, the force equation gives the electric field necessary to preserve charge neutrality:

$$\mathbf{E} = -\frac{\nabla p_e}{ne} - \frac{\mathbf{u} \times \mathbf{B}}{c}. \tag{12.12}$$

Here $\mathbf{u}$ is the plasma flow velocity, p_e the electron pressure, and $\mathbf{B}$ the magnetic field. Substitution of $\mathbf{E}$ into Faraday's law then gives

$$\frac{1}{c}\frac{\partial \mathbf{B}}{\partial t} = \nabla \times \left(\frac{\mathbf{u} \times \mathbf{B}}{c} + \frac{\nabla p_e}{ne} \right). \tag{12.13}$$

Equation (12.13) shows that a magnetic field is generated whenever the condition $\nabla \times (\nabla p_e / n) \neq 0$ holds i.e., when $\nabla n \times \nabla \theta_e \neq 0$.

To show that a sizable inhibition is possible via these fields, consider here a crude order-of-magnitude estimate. Setting $\partial B/\partial t = 0$ and denoting all gradient lengths by some L, we find with Eq. (12.13) that $|B| \approx (c/u)(\theta_e/eL)$. The energy flux carried by electrons across the field can be estimated as $I_e \approx n\theta_e D/L$, where D is the diffusion coefficient. Since the B fields can be quite inhomogeneous, we conservatively take Bohm-like diffusion ($D \approx \omega_{ce} r_{ce}^2$, where ω_{ce} is the electron cyclotron frequency and r_{ce} the gyroradius). Combining these estimates then gives $I_e \approx n\theta_e u$, that is, the characteristic energy flow speed is the plasma flow velocity which is typically of the order of the ion sound velocity.

For quantitative results, it is essential to address many important questions — the size and extent of the self-generated fields, how electrons diffuse across them and other mechanisms for their generation. The B fields can also be generated by other mechanisms such as instabilities produced by the anisotropic heated electron distribution or by velocity anisotropies associated with the heat flow. The resulting B fields can be quite inhomogeneous, and so their influence [19] on the electrons is itself a very rich topic.

Lastly the transport coefficient can also be modified by ion turbulence in the plasma. The most commonly invoked mechanism for producing this turbulence is the ion-acoustic drift instability driven by the heat flow [20–22]. The basic idea is very simple. The electron distribution function carrying a heat flux Q is skewed, as shown by Eq. (12.8). In particular, the low-energy electrons have a drift $V_d \sim Q/n\theta_e$ relative to the ions. Physically this drift is produced by the self-consistent electric field necessary to draw a return current to compensate for the flow of the hotter electrons which carry the heat flow. When this drift exceeds the threshold for the ion-acoustic drift instability, ion-acoustic waves are driven unstable. It has been hypothesized that the ion turbulence then so effectively scatters the electrons that the heat flow is locked into a value near the instability threshold. In other words, $Q_{max} \approx n\theta_e v_s$, when $Z\theta_e \gg \theta_i$. Here v_s is the ion sound velocity and θ_i is the ion temperature. There has been considerable controversy over whether the ion-acoustic turbulence can be this effective, particularly for electrons of very high energy. Computer simulations [23] have suggested that the ion turbulence does not strongly limit the heat flux.

12.4   HEAT TRANSPORT IN LASER-IRRADIATED TARGETS

Theoretical description of large electron heat fluxes in laser-irradiated targets is clearly a challenging problem. Let's conclude this chapter with a very brief synopsis of the experimental feedback. In laser-irradiated targets, the heat transport has been inferred from a variety of different measurements, including x-ray images of the heated plasma, the ratio of the energy in fast and slow ion expansion, the implosion and mass ablation efficiencies, the density profile in the underdense plasma, and the burn-through rate of thin films and layered targets.

As an example, let's consider experiments [24] in which Al disks coated with a layer of CH were irradiated with a 100 ps pulse of 1.06μ laser light with a peak intensity of 10^{15}W/cm^2. The thickness of the CH layer was varied, and the x-ray emission at energies between 1 and 3 keV was measured as a monitor of the energy transported to the Al substrate. As shown in Fig. 12.1, a rather thin layer of CH led to an abrupt decrease in the x-ray emission, indicating poor electron transport. Calculations of electron energy transport model this data by using a flux limit of $f \sim 0.01$.

A similar inhibition of the heat transport ($f \sim 0.01$–0.04) has been inferred from many other experiments [25–29] on disk targets. The transport inhibition appears to decrease as the intensity and/or the wavelength of the light is reduced. Furthermore, in at least some experiments [30–36] in which spherical targets are rather uniformly irradiated, the electron energy transport appears to saturate at a value close to that expected from numerical calculations of the Fokker-Planck equation (i.e., $Q \sim 0.1\, n\,\theta_e v_e$). This suggests that B field generation and/or lateral energy transport may be playing a significant role in the disk experiments. Many of the recent transport experiments and calculations are reviewed in Ref. 15.

Although much remains to be understood, there has clearly been significant progress in both characterizing and understanding electron energy transport in laser irradiated targets. More experiments are needed to clarify the dependence of the heat transport on such critical features as target geometry, uniformity of irradiation, and laser light wavelength and intensity. Improved measurements of the underdense plasma conditions are also needed. These conditions are an important indicator of the transport and directly influence the coupling processes, which serve as source terms driving the energy transport.

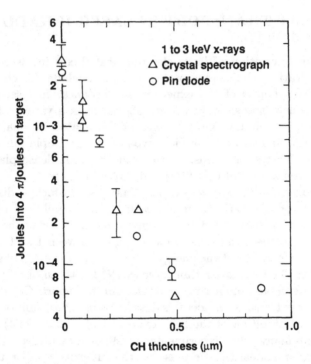

Figure 12.1   The fraction of incident laser energy converted in Al line radiation (Δ) and into 1–3Kev x-rays (o) as a function of the thickness of the CH overlayer. See *Young et al.*, (1977).

References

1. Spitzer, L. and R. Harm, Transport phenomena in a completely ionized gas, *Phys. Rev.* **89**, 977 (1953).

2. I. P. Shkarovsky, T. W. Johnston, and M. P. Backynski, *The Particle Kinetics of Plasmas*. Addison-Wesley, Reading, Mass., 1966.

3. Zimmerman, G. P. and W. L. Kruer, Numerical simulation of laser-irradiated fusion, *Comments Plasma Phys. Controlled Fusion* **2**, 51 (1975).

4. Gray, D. R. and J. D. Kilkenny, The measurement of ion acoustic turbulence and reduced thermal conductivity caused by a large temperature gradient in a laser heated plasma, *Plasma Physics* **22**, 81 (1980).

5. Shvarts, D., J. Delettrez, R. L. McCrory, and C. P. Verdon, Self-consistent reduction of the Spitzer-Härm electron thermal heat flux in steep temperature gradients in laser-produced plasmas, *Phys. Rev. Lett.* **47**, 247 (1981).

6. Bell, A. R., R. G. Evans, and D. J. Nicholas, Electron energy transport in steep temperature gradients in laser-produced plasmas, *Phys. Rev. Lett.* **46**, 243 (1981).

7. Mason, R. J., Apparent and real thermal inhibition in laser-produced plasmas, *Phys. Rev. Lett.* **47**, 652 (1981);

8. Khan, S. A. and T. D. Rognlein, Thermal heat flux for arbitrary collisionality, *Phys. Fluids* **24**, 1442 (1981).

9. Matte, J. P. and J. Virmont, Electron heat transport down steep temperature gradients, *Phys. Rev. Lett.* **49**, 1936 (1982).

10. Matte, J. P., T. W. Johnston, J. Delettrez, and R. L. McCrory, Electron heat transport with inverse bremsstrahlung and ion motion, *Phys. Rev. Lett.* **53**, 1461 (1984).

11. Luciani, J. F., P. Mora, and J. Virmont, Nonlocal heat transport due to steep temperature gradients, *Phys. Rev. Lett.* **51**, 1664 (1983).

12. Luciani, J. F., P. Mora, and R. Pellat, Quasistatic heat front and delocalized heat flux, *Phys. Fluids* **28**, 835 (1985).

13. Albritton, J. R., Laser absorption and heat transport by non-Maxwell-Boltzmann electron distributions, *Phys. Rev. Lett.* **50**, 2078 (1983).

14. Albritton, J. R., E. A. Williams, I. B. Bernstein, and K. P. Swartz, Nonlocal electron heat transport by not quite Maxwell-Boltzmann distributions, *Phys. Rev. Lett.* **57**, 1887 (1986).

15. For a review of calculations and experiments on transport, see J. Delettrez, Thermal electron transport in direct-drive laser fusion, *Can. J. Phys.* **64**, 932 (1986).

16. Haines, M. G., Magnetic-field generation in laser fusion and hot-electron transport, *Can. J. Phys.* **64**, 912 (1986).

17. Stamper, J. A. *et al.*, Spontaneous magnetic fields in laser-produced plasmas, *Phys. Rev. Lett.* **26**, 1012 (1971).

18. Forslund, D. W. and J. U. Brackbill, Magnetic-field-induced surface transport on laser-irradiated foils, *Phys. Rev. Lett.* **48**, 1614 (1982).

19. Max, C. E., W. M. Manheimer, and J. J. Thomson, Enhanced transport across laser generated magnetic fields, *Phys. Fluids* **21**, 128 (1978).

20. Forslund, D. W., Instabilities associated with heat conduction in the solar wind and their consequences, *J. Geophys. Res.* **75**, 17 (1970).

21. Bickerton, R. J., Thermal conduction limitations in laser fusion, *Nucl. Fusion* **13**, 457 (1973).

22. Manheimer, W. M., Energy flux limitation by ion acoustic turbulence in laser fusion schemes, *Phys. Fluids* **20**, 265 (1977).

23. Lindman, E. L., Absorption and transport in laser plasmas, *Journal de Physique* **38**, Colloque C6, 9 (1977).

24. Young, F. C. *et al.*, Laser-produced-plasma energy transport through plastic films, *Appl. Phys. Lett.* **30**, 45 (1977).

25. Yaakobi, B. and T. C. Bristow, Measurement of reduced thermal conduction in (layered) laser-target experiments, *Phys. Rev. Lett.* **14**, 350 (1977).

26. Malone, R. C., R. L. McCrory, and R. L. Morse, Indications of strongly flux-limited electron thermal conduction in laser-target experiments, *Phys. Rev. Lett.* **34**, 721 (1975).

27. Campbell, E. M., R. R. Johnson, F. J. Mayer, L. V. Powers, and D. C. Slater, Fast-ion generation by ion-acoustic turbulence in spherical laser plasmas, *Phys. Rev. Lett.* **39**, 274 (1977).

28. Yaakobi, B. *et al.*, Characteristics of target interaction with high power UV laser radiation, *Opt. Commun.* **39**, 175 (1981).

29. Mead, W. C. *et al.*, Characteristics of lateral and axial transport in laser irradiations of layered-disk targets at 1.06μm and 0.35μm wavelengths, *Phys. Fluids* **27**, 1301 (1984).

30. Goldsack, T. J. *et al.*, Evidence for large heat fluxes from the mass ablation rate of laser-irradiated spherical targets, *Phys. Fluids* **25**, 1634 (1982).

31. Yaakobi, B. *et al.*, Thermal transport measurements in 1.05μm laser irradiation of spherical targets, *Phys. Fluids* **27**, 516 (1984).

32. Tarvin, J. A., W. B. Fechner, J. T. Larsen, P. D. Rockett, and D. C. Slater, Mass-ablation rates in a spherical laser-produced plasma, *Phys. Rev. Lett.* **51**, 1355 (1983).

33. Fechner, W. B., C. L. Shephard, G. E. Busch, R. J. Schroeder, and J. A. Tarvin, Analysis of plasma density profiles and thermal transport in laser-irradiated spherical targets, *Phys. Fluids* **27**, 1552 (1984).

34. Haver, A. *et al.*, Measurement and analysis of near-classical thermal transport in one-micron laser-irradiated spherical plasmas, *Phys. Rev. Lett.* **53**, 2563 (1984).

35. Yaakobi, B. *et al.*, Thermal transport measurements in six-beam, ultraviolet irradiation of spherical targets, *J. Appl. Phys.* **57**, 4354 (1985).

36. Jaanimagi, P. A., J. Delettrez, B. L. Henke, and M. C. Richardson, Temporal dependence of the mass-ablation rate in UV-laser irradiated spherical targets, *Phys. Rev. A* **34**, 1322 (1986).

CHAPTER 13

Laser
Plasma
Experiments

In previous chapters, we have discussed a variety of mechanisms for laser plasma coupling, ranging from collisional absorption to excitation of many different instabilities. Figure 13.1 illustrates the rich variety of coupling processes as a function of the plasma density. Near the critical density (n_{cr}), we have resonance absorption and instabilities leading to the excitation of electron and ion waves. Near $n_{cr}/4$, we have the $2\omega_{pe}$ instability. The Raman instability operates for densities $\lesssim n_{cr}/4$. The Brillouin and filamentation instabilities take place throughout the underdense plasma, as does inverse bremsstrahlung absorption. Throughout the underdense plasma there can be self-generated magnetic fields or elevated levels of ion turbulence driven by a variety of processes associated with the plasma heating and expansion.

The mix of coupling processes depends on the intensity, wavelength, and beam quality of the laser light and upon the gradient lengths, plasma composition, and other plasma conditions. In turn, these depend on the mix of coupling processes. An understanding of this coupled nonlinear

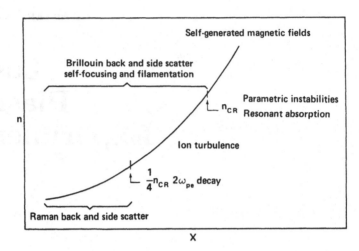

Figure 13.1   A schematic of the density profile in the underdense plasma illustrating the many processes which affect laser plasma coupling.

problem requires close collaboration between theory, computer simulation and experiments. Although the understanding is far from complete, calculations have at least qualitatively predicted many important plasma effects observed in experiments. These include steepening of the density profile, hot electron generation, resonance absorption, stimulated Raman and Brillouin scattering, and favorable wavelength scaling. In this final chapter, let us briefly consider some of the experimental evidence for various laser plasma processes.

The characteristic size (L) of the underdense plasma is a very useful parameter to consider when discussing experiments. If L is small $(L/\lambda_0 \lesssim \mathcal{O}(10)$, where λ_0 is the free space wavelength of the laser light), then many of the coupling processes are either before threshold or weakly occurring, and we are primarily concerned with how light is absorbed near the critical density surface. On the other hand, if there is an extensive region of underdense plasma $[L/\lambda_0 \gtrsim \mathcal{O}(100)]$, theory indicates that processes such as Brillouin and Raman scattering, filamentation and inverse bremsstrahlung can begin to play a significant role.

We can estimate the size of the underdense plasma in laser-irradiated targets as the minimum of $c^*\tau/2$ or R where c^* is a typical plasma expansion velocity, τ is the pulse length of the laser light, and R is the focal spot radius. To give some feeling for the numbers, let us take an expansion velocity of 3×10^7 cm/sec, which is approximately the ion sound speed in a hydrogen plasma with an electron temperature of 1 keV. Then

$$\frac{L}{\lambda_0} \sim \frac{\min\left[1.5 \times 10^2\,\tau(\text{ns}),\ R(\mu)\right]}{\lambda_0(\mu)}\,,$$

where 'min' denotes minimum, τ is expressed in nanoseconds, and R and λ_0 are expressed in micrometers. Hence experiments with 1.06μm light and pulse lengths of $\lesssim$ 30ps have rather small underdense plasmas, whereas experiments with pulse lengths $\gtrsim$ 1ns and large focal spot have large underdense plasmas. Note also the scaling as τ/λ_0. With this distinction in mind, we will first examine some short-pulse-length experiments (with small underdense plasmas) and then consider some longer pulse-length ones.

13.1 DENSITY PROFILE STEEPENING

As we discussed in Chapter 10, calculations show a pronounced steepening of the density profile near the critical density. This steepening is important because the scale length near the critical density affects the mix of absorption processes and the heated electron temperatures. This profile steepening has been confirmed by interferometric measurements [1–3] of the density of a laser-heated plasma. In one experiment [1] a 41μm diameter glass microballoon was irradiated with a 30ps, 1.06μm laser pulse at an intensity of 3×10^{14}W/cm^2. An interferogram was taken with a 15ps, 0.26μm probe beam and Abel-inverted to determine the axial electron density profile plotted in Fig. 13.2. In both experiment and simulations the profile is steepened to an upper density n_u that is roughly determined by pressure balance:

$$n_u \simeq n_{\text{cr}}\left[1 + \left(\frac{v_{\text{os}}}{v_e}\right)^2\right],$$

where v_{os} is the oscillation velocity of an electron in the laser light field, v_e is the electron thermal velocity and n_{cr} is the critical density. The profile is steepened down to a lower density that is determined by how the light pressure and localized heating dams the plasma flow. This lower density typically appears to be somewhat less in experiments than in the particle simulations, perhaps because of energy-transport inhibition.

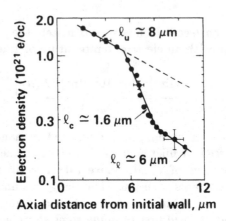

Figure 13.2   A plasma density profile in a laser-irradiated target measured by interferometry using a 0.26μm light pulse. See *Attwood et al.* (1978).

Calculations with a focused light beam [4] show an additional, related effect: cratering of the critical density surface. Physically, the density surface is preferentially pushed in where the light intensity is greatest. This effect has also been observed in experiments. Figure 13.3 shows an Abel inverted transverse density contour measured in an experiment [1] in which a disk target was irradiated with 1.06μm light with an intensity of $\simeq 3 \times 10^{14}$ W/cm^2. The density cavity has a transverse scale length approximately equal to that of the incident light beam. A smaller scale rippling of the critical density surface has also been inferred in experiments using higher intensity light. These ripples are probably due to hot spots in the incident light beam and/or to a critical surface instability found in computer simulations [5–7].

13.2   ABSORPTION OF INTENSE, SHORT PULSE-LENGTH LIGHT

Many important features of the absorption measured in experiments with short pulse length, high intensity laser light can be understood in terms of resonance absorption in a steepened density profile. In such experiments, the underdense plasma has both a high temperature and a small spatial extent. Because collisional absorption varies as $\theta_e^{-3/2}$ (θ_e is the

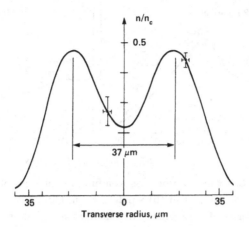

Figure 13.3 A transverse density profile measured by interferometry. See *Attwood et al.*, (1978).

electron temperature) and as the scale length of the plasma, it is relatively weak in these experiments. However, as discussed in Chapter 4, computer simulations showed that there would still be a sizeable absorption due principally to resonance absorption with some additional absorption due to nonlinearly-generated ion fluctuations. These early simulation results assumed plane waves incident onto a plasma slab. In practice, a focussed light beam (say, with hot spots) both craters and ripples the critical density surface as discussed in the previous section. These surface ripples average the absorption over angle as well as change part of the p-polarized light into s-polarized light and vice versa [8–10]. A simple theory [8] was used to extend the ideal simulation results to crudely include this additional critical surface rippling. The result for the absorption as a function of polarization and angle of incidence is shown by the black line in Fig. 13.4.

The absorption has been measured in detail in numerous experiments [11–17]. In some of these experiments [11], plastic disks were irradiated with about 30ps pulses of $1.06\mu m$ light with an intensity of 10^{15} - 10^{16}W/cm^2. The measured absorption, denoted by the circles in Fig. 13.4, was both polarization-dependent and broad in angle. The absorption of p-polarized light peaked at approximately the predicted angle, and the absorption of s-polarized light monotonically decreased with the angle of incidence. The magnitude of the absorption was also in reasonable agree-

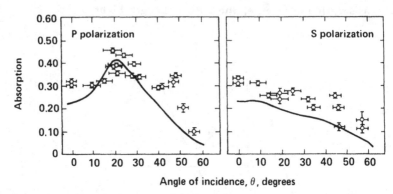

Figure 13.4   Laser light absorption as a function of angle of incidence, shown here for (a) p-polarization and (b) s-polarization. Circles denote the absorption measured in a series of experiments in which plastic disks were irradiated with 30ps pulses of 1.06μm light. See *Manes et al.*, (1977).

ment. The principal discrepancy is an additional 10–15% absorption, rather independent of angle of incidence and polarization. This additional absorption may be due to a number of effects such as inverse bremsstrahlung, critical surface rippling, or self-generated magnetic fields. At very high intensities ($I\lambda_0^2 \gtrsim 10^{17}$W-$\mu^2$/cm^2), the measured fractional absorption increases significantly [16]. This effect has been attributed to the development of a highly turbulent critical surface.

13.3   HEATED ELECTRON TEMPERATURES

The heated velocity distribution is a very important feature of the absorption process. The light primarily heats electrons, since their motion in the oscillating fields is much larger than that of the massive ions. As discussed in the Chapter 9, absorption via plasma waves in general leads to high energy tails on the electron distribution function. Physically this is because plasma waves tend to preferentially heat the faster (more nearly resonant) electrons. The simulations of resonance absorption discussed in Chapter 10 predict that a roughly two-temperature distribution will result [18,19]. The lower temperature is that typical of electrons streaming into the absorption region and is determined by how the heat transports to higher density. The hot temperature is that characteristic of the electrons heated by resonance absorption, which in the simulations are found to have a quasi-Maxwellian velocity distribution.

The generation of an electron distribution with at least two character-istic temperatures is supported by measurements of the x-ray spectrum in many different experiments using short-pulse-length laser light. As an ex-ample, Fig. 13.5 shows the x-ray spectrum observed in an experiment [20] in which a plastic disk was irradiated with a 80ps pulse of 1.06μm light at a peak intensity of $\simeq 2 \times 10^{15}$W/cm^2. The low energy x-rays indicate an electron temperature of $\simeq$ 700eV, and the high energy x-rays a tem-perature of $\simeq$ 8 keV. A third, higher temperature component has been observed in experiments with small underdense plasmas for sufficiently intense irradiation [16,17].

Figure 13.6 shows the heated electron temperature inferred from the high energy x-rays in short pulse experiments over a wide intensity regime. The various symbols with error bars represent data from a series of exper-iments [21] in which disks or microballoons were irradiated with 1.06μm light with pulse lengths in the range of about 50–200ps. The open x's are values of the resonantly-heated electron temperature calculated in a se-ries of two-dimensional simulations [18] of resonance absorption. Both the magnitude and intensity scaling of the heated electron temperatures are in reasonable agreement, especially in view of the fact that the simulations are quite ideal and do not include the space-and time-averaging inherent in the experiments. Experiments using laser light with other wavelengths

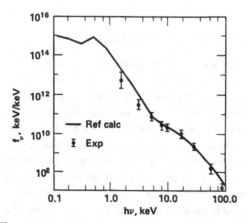

Figure 13.5   The x-ray spectrum measured in an experiment in which a plastic disk was irradiated by a 80ps pulse of 1.06μm light with an intensity of about 2×10^{15} W/cm^2. See *Haas et al.*, (1977).

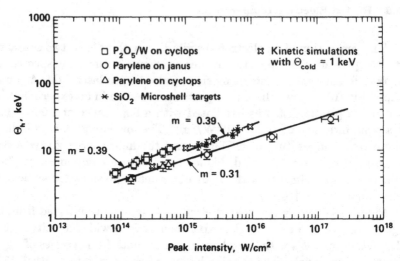

Figure 13.6   The heated electron temperature as a function of incident laser light intensity. The symbols with error bars are values inferred from the high energy x-rays in experiments in which disks or microballoons were irradiated with 1.06μm light. The open x's are values calculated in a series of two-dimensional simulations. See *Manes et al.*, (1977).

have given quite similiar results. One simply has to scale the intensity as $I\lambda_0^2$ (where λ_0 is the free-space wavelength of the light), as theoretically expected.

13.4   BRILLOUIN SCATTERING

We have been discussing short-pulse-length experiments which are characterized by a small region of plasma with density less than n_{cr}. These experiments have been typical of the exploding-pusher target experiments carried out in the early days of the laser fusion program. Laser plasma coupling is more complex in long-pulse-length experiments with larger regions of underdense plasma. Such experiments are more characteristic of the ablative compressions needed to compress fuel to a high density to achieve high gain [22]. In the longer scale length plasmas, effects such as collisional absorption, Brillouin and Raman scattering, and filamentation can all play a much more important role. Let's now discuss experiments on some of these processes.

Experiments [23–37] with longer scale length plasmas and 1.06μm light show that significant Brillouin scattering is possible. In these experiments, sizeable underdense plasmas ($L/\lambda_0 > 30$) were formed in various ways: by using a prepulse, a long pulse length in the sense of τ/λ_0, or a preformed plasma. Let's consider a specific example in which a larger underdense plasma was created by use of a prepulse. As shown in Fig. 13.7, the addition of a prepulse about 2ns prior to a 75ps main pulse was found to increase the backscatter of the main pulse which was normally incident onto a CH slab [23]. As the prepulse energy was increased, the fraction of the main pulse which was back-reflected increased from $\simeq 15\%$ to $\simeq 40\%$ and the net absorption decreased from $\simeq 50\%$ to $\simeq 20\%$. For a fixed ratio of prepulse energy to main pulse energy, the backscattered light increased with the intensity of the main pulse and was rather insensitive to the angle at which the targets were tilted. In addition, it was shown that the light rays retraced their path. All these features are as expected if the light reflection is due to the Brillouin instability in the underdense plasma.

Experiments have also shown evidence for Brillouin sidescatter. This instability preferentially scatters light out of the plane of polarization. As shown in Fig. 13.8, the sideward scattering out of the plane of polarization has been observed to greatly exceed that into the plane of polarization in experiments with large focal spots [27]. Typically, the polarization-dependent, sideward scattering observed to date has been $\lesssim 30\%$ of the incident laser energy.

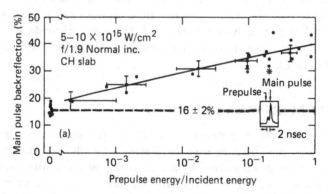

Figure 13.7   The fraction of the main pulse energy which was backscattered into an $f/1.9$ lens versus the fraction of the energy into a prepulse. See *Ripin et al.*, (1977).

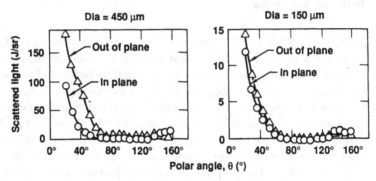

Figure 13.8   The angular distribution of the light scattered from a Au disk irradiated with 1.06μm light focussed to an intensity of about 3×10^{14} W/cm^2. The focal spot diameters are 150μm and 450μm. See *Rosen et al.*, (1979).

Quite detailed studies of the nonlinear aspects of Brillouin scattering have been carried out in microwave experiments with low density plasmas [38,39]. In some of these experiments, the amplitudes of both the driven ion wave and its harmonic were measured via Thomson scattering of millimeter waves. In addition, a heated ion tail was observed and its density measured. These experiments emphasize that ion tail formation is a very potent saturation mechanism.

The behavior of Brillouin scattering in laser experiments needs further study. Identification of the scattering by its frequency spectrum is often uncertain due to Doppler shifts in the expanding plasma. Often the observed scattered light with frequency near ω_0 does not appear to exhibit any well-defined growth or saturation [32]. In experiments to date with overdense targets and short wavelength light, the backscattering is quite modest ($\lesssim 10\%$).

13.5   RAMAN SCATTERING

There are numerous experiments [40–53] on Raman scattering in laser-irradiated plasmas. This process is relatively simple to identify since the scattered light is down-shifted by an electron plasma frequency. In experiments with a large region of underdense plasma, the energy in Raman-scattered light has been measured to be as large as 10–20% of the incident laser energy. In addition, the expected correlation with hot electron generation has been observed. Let us briefly mention several of these experiments.

In early experiments which showed a significant level of Raman scattering [47], a large region of underdense plasma was formed by irradiating a thin CH foil (7000Å thick) with a 900ps, 3kJ pulse of 1.064μm light. Computer calculations predicted this foil to go underdense and expand through a density of a few tenths of the critical density somewhat before the peak of the laser pulse. When this happens, the laser light is propagating through a relatively flat region of underdense plasma. If we simply use the radius of the focal spot as a measure of the distance over which the plasma density is reasonably flat, the characteristic plasma size is of order 200μm, which is sufficient to produce sizeable Raman scatter. Indeed, about 10% of the light was observed to be Raman scattered. Figure 13.9 shows the measured angular distribution of the spectrally integrated Raman scattered light. Most of the scattering was into the rear hemisphere, but even a small fraction of the incident light was scattered into the forward hemisphere.

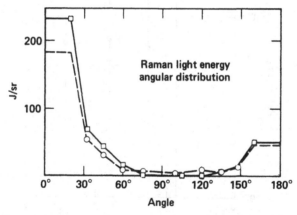

Figure 13.9   The angular distribution of the spectrally-integrated Raman scattered light measured in experiments in which thin foils were irradiated with 1.06μm light. The different symbols denote two separate experiments. See *Phillion et al.*, (1982).

Subsequent experiments [49] with thin foils irradiated with 0.53μ and 0.26μ light have shown comparable levels ofRaman scattering. As expected on the basis of collisional damping, the scattering was found to decrease dramatically for Au foils irradiated with 0.26μ light. In other thin foil experiments using 0.53μ light, both the up- and down-shifted components expected from Raman forward scattering were observed [50].

In experiments [51] at the University of Alberta, Raman backscatter was observed when a rather uniform low density plasma in a solenoid was irradiated with CO_2 laser light. The plasma density was about $1/40$ of the critical density, the background electron temperature about 80 eV, and the interaction length about 3mm, as estimated from the depth of focus of the laser light. As shown in Fig. 13.10, back reflection due to the Raman instability was observed to onset at an intensity of $\sim 4 \times 10^{10}$ W/cm^2, which was calculated to be the expected threshold intensity. When the intensity was further increased, the reflectivity from this rather low density plasma saturated at a value of about 0.7%.

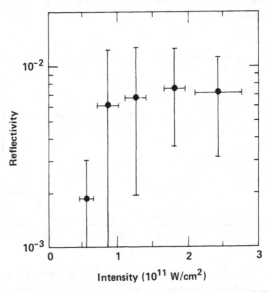

Figure 13.10   The back reflection due to the Raman instability measured in experiments in which a low density plasma in a solenoid was irradiated with 10.6μm light. See *Offenberger et al.*, (1982).

Raman reflected light has also been observed in experiments with rather inhomogeneous plasma blowing off from an overdense target. The spectra measured in such experiments [45,46] have indicated Raman scattering from the region near $n_{cr}/4$ as well as from the plasma at lower densities down to about $0.05–0.1n_{cr}$. The level of the scattered light in these experiments is often quite low: in the range of $10^{-6}–10^{-4}$ of the incident light. However, this level has been found to increase rapidly as the focal spot size and pulse length is increased, leading to more gentle gradients. For example, in experiments [47,52] in which thick disk targets were irradiated with 1.06μ or 0.53μ light, up to several percent of the incident laser light was observed to be Raman scattered.

Finally a correlation of Raman scattering with hot electron generation has been observed in experiments [52] in which Au disks were irradiated with 1ns pulses of $0.53\mu m$ light. In these experiments, the laser energy varied from 0.5–4.0kJ and the nominal intensity from about $10^{14}–2\times10^{16}$W/cm^2. The slope of the high energy x-rays indicates hot electrons with a temperature of about 30keV. Figure 13.12 shows the fraction of the laser energy deposited into hot electrons as inferred from the level of the hard x-rays versus the measured fraction of the laser energy in

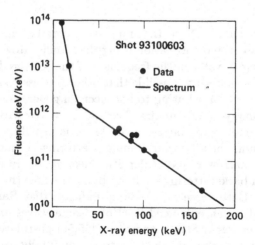

Figure 13.11   Hard x-ray spectrum from Au disk irradiated with a 3.6kJ, 1ns pulse of $0.53\mu m$ light focussed onto $740\mu m$ spot. See *Drake et al.*, (1984).

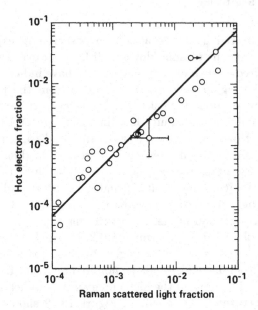

Figure 13.12   The fraction of the laser energy absorbed into hot electrons versus the fraction in Raman-scattered light in Au disks irradiated by 1ns pulses of 0.53μm light. See *Drake et al.*, (1984).

Raman-scattered light. Note the impressive correlation. The solid line represents the expected correlation using the Manley-Rowe relations with the measured mean value of the frequency of the scattered light. Because of the error bars, it is quite possible that other processes such as the $2\omega_{pe}$ instability are also contributing to hot electron generation.

Although many trends in the observations agree with expectations, there are also challenging puzzles [54]. There is usually a gap in the frequency spectrum, showing that Raman scattering is much weaker than expected for a narrow range of densities near $n_{cr}/4$. In addition, a low level of Raman backscattering is often observed below the nominal intensity threshold. Both these puzzles may indicate that Raman scattering is being seeded by an enhanced level of plasma waves excited by other processes. For example *Simon and Short* [55,56] postulate that bursts of hot electrons due to the $2\omega_{pe}$ instability preferentially excite the plasma waves in the lower density region. Below the Raman instability threshold, we would then have ordinary Thomson scattering from enhanced fluctuations. Above threshold, the instability grows from the enhanced levels.

13.6   OTHER PLASMA PROCESSES

There is also experimental evidence for many other plasma processes in laser-produced plasmas. Electron plasma waves due to the $2\omega_{pe}$ instability have been directly observed by Thomson scattering in experiments with 10.6μm light. The growth rate, the local density profile steepening, and the generation of ion waves in the nonlinear state have been measured in some detail [57–59]. A useful signature of this instability is emission near $3\omega_0/2$, which arises from the coupling of the incident and reflected light wave with a plasma wave near $n_{cr}/4$. Unfortunately the level of the instability is difficult to estimate from this signal, since the emission only indirectly indicates the level of part of the spectrum of driven plasma waves.

The $3\omega_0/2$ emission is frequently diagnosed in laser plasma experiments [60–62]. For example, in some experiments at the University of Rochester [62], CH spheres were irradiated with a 600–700ps pulse of 0.35μm light. The $3\omega_0/2$ emission was observed to onset at an intensity of about 2×10^{14}W/cm^2, the estimated threshold intensity of the $2\omega_{pe}$ instability. The level of the emission increased with the intensity of irradiation but then saturated at rather a low level when the intensity was about 3×10^{14} W/cm^2. Hard x-rays indicating suprathermal electrons with a temperature of about 35keV were observed to be correlated with the $3\omega_0/2$ emission. The inferred fraction of the laser energy in these suprathermal electrons saturated at a low value of about 10^{-4} of the incident energy in these experiments with $L/\lambda_0 \lesssim 150$.

Filamentation of laser light is perhaps the least characterized of the plasma processes we have discussed. Much of the evidence is rather indirect: inferrences from structure [63–65] in x-ray pictures of the heated plasma or in images of the back-reflected light. Filamentation has also been inferred from the angular distribution [45] of the half-harmonic light or from frequency shifts [66] in the reflected light. Filamentary structures have been directly observed by using optical shadowgraphy [67], by imaging the second harmonic emission [68], and by Thomson scattering from electron plasma waves generated in the walls of the filament [69].

Parametric instabilities near the critical density have been inferred from frequency shifts in the second harmonic emission [70–72] as well as from Thomson scattering measurements [73] of ion acoustic waves induced by a 10.6μm laser. The unstable waves and the plasma heating have been measured in some detail in microwave experiments with low density plasmas [74–77]. Excitation of waves near the critical density has also been studied extensively in ionospheric heating experiments [78,79]. In addition, self-generated magnetic fields with values up to 10^6 Gauss have been measured in laser experiments using Faraday rotation of a probe beam [80,81]. Since we've already considered a number of different plasma processes, we will now proceed to the important topic of wavelength scaling.

13.7   WAVELENGTH SCALING OF LASER PLASMA COUPLING

As both calculations and experiments have amply demonstrated, laser plasma coupling can be influenced by a rich variety of collective plasma effects. Many of these collective processes either decrease the absorption or give absorption into a tail of very energetic electrons. The advantages of enhancing collisional absorption and reducing collective effects have placed a premium on the use of short wavelength laser light [82,83].

When the laser wavelength (λ_0) is decreased, the light wave penetrates to higher density plasma since the critical density increases as λ_0^{-2}. For a given absorbed intensity, the heated plasma is both denser and lower in temperature and hence much more collisional. In addition to being reduced by this greater collisionality, the collective processes are more weakly driven by short wavelength light. For a given intensity, the oscillation velocity of an electron in the light wave is proportional to the laser wavelength.

Many experiments [84–88] with 0.53μm, 0.35μm and 0.26μm light have demonstrated that important features of the coupling improve as the wavelength decreases. Figure 13.13 shows a compendium [89] of the absorption as a function of intensity measured in a variety of experiments using laser light with wavelengths ranging from 1.05μm to 0.26μm. In these experiments on CH targets, the pulse lengths varied from 100ps to 1ns, but the focal spot size was typically rather small. Note the dramatic increase in absorption as the wavelength decreases, as expected since inverse bremsstrahlung depends strongly on wavelength. A very strong decrease in hot electron generation has also been shown in such experiments. Figure 13.14 shows x-ray spectra measured in experiments [90] in which a 600ps pulse of light was focused to an intensity of about 3×10^{14} W/cm^2 onto an Al disk. The level of the high energy x-rays decreased by several orders of magnitude as the wavelength of the light was changed from

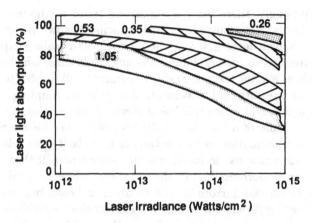

Figure 13.13   The absorption versus intensity measured using laser light with wavelengths ranging from $1.05\mu m$ to $0.26\mu m$. See *Ripin and Kruer*, (1986).

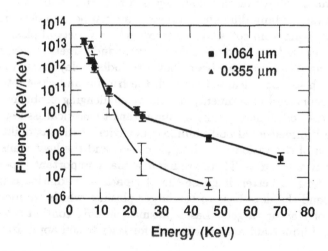

Figure 13.14   The x-ray spectra from an Al disk irradiated with $\sim 600ps$ pulses of $1.064\mu m$ and $0.355\mu m$ light. The absorbed intensity was about 1.5×10^{14} W/cm^2. See *Campbell*, (1984).

$1.06\mu m$ to $0.355\mu m$, illustrating a significant reduction in collective plasma interactions.

Not surprisingly, there has been a strong trend to the use of short-wavelength lasers. Most large laser fusion facilities now operate or plan to operate at wavelengths $\lesssim 0.5\mu$. These wavelength scaling experiments are being extended to include long-scale-length plasmas, which are more characteristic of reactor targets. Such experiments will be able to quantify the regimes of intensity and wavelength for optimum coupling.

Finally, increased attention is being focussed on techniques to smooth the intensity profile of a laser beam [91–94]. Smoother beams will allow more uniform illumination of fusion targets and clarify the role of beam structure in experiments. Induced spatial incoherence (ISI) is one technique for beam smoothing [91]. In the usual embodiment of ISI, a pair of reflecting, echelon-like mirrors is used to divide a broad bandwidth laser beam into many independent beamlets. The echelons introduce time delays between the beamlets which are longer than the laser coherence time. The beamlets are overlapped onto a target, producing a very smooth intensity profile when averaged over time scales long compared to the coherence time. Experiments [94] with induced spatial incoherence are showing that the use of a smoother beam further improves the coupling of short wavelength laser light with targets.

In summary, laser plasma coupling is a rich and challenging area of applied physics. Many different processes can compete to determine the coupling, ranging from collisional absorption to a variety of plasma instabilities. As lasers have become more energetic and targets larger, more of these plasma processes have been shown to indeed play a role in experiments. Much has been learned through the fruitful and close interaction between theory and experiment, and many challenging problems remain to be understood. There are many important theoretical issues, including the multidimensional and nonlinear behavior of the instabilities, the competition of the various coupling processes, and the heat transport in strongly driven plasmas. There are likewise many important experiments yet to be done to better diagnose the plasma and irradiation conditions and to extend the data base to the larger plasmas which are more characteristic of reactor targets. Laser plasma coupling continues to be an exciting and important area of research for laser fusion applications.

References

1. Attwood, D. T., D. W. Sweeney, J. M. Auerbach, and P. H. Y. Lee, Interferometric confirmation of radiation-pressure effects in laser-plasma interactions, *Phys. Rev. Lett.* **40**, 184 (1978).

2. Fedosejevs, R., L. V. Tomov, N. H. Burnett, G. D. Enright, and M. C. Richardson, Self-steepening of the density profile of a CO_2-laser-produced plasma, *Phys. Rev. Lett.* **39**, 932 (1977).

3. Fechner, W. B., C. L. Shepard, Gar. E. Busch, R. J. Schroeder, and J. A. Tarvin, Analysis of plasma density profiles and thermal transport in laser-irradiated spherical targets, *Phys. Fluids* **27**, 1552 (1984).

4. Randall, C. J. and J. DeGroot, Effect of crater formation on the absorption of focused laser light, *Phys. Rev. Lett.* **42**, 179 (1979).

5. Estabrook, K. G., Critical surface bubbles and corrugations and their implication to laser fusion, *Phys. Fluids* **19**, 1733 (1976).

6. Nishimura, H. *et al.*, Resonance absorption and surface instability at a critical density surface of a plasma irradiated by a CO_2 laser, *Plasma Phys.* **21**, 69 (1980).

7. Bezzerides, B. *et al.*, Recent developments in understanding the physics of laser-produced plasmas; in *Proceedings of the Sixth International Conference on Plasma Physics and Controlled Fusion Research*, Vol. 1, p. 123. International Atomic Energy Agency, Vienna, 1977.

8. Thomson, J. J., W. L. Kruer, A. B. Langdon, C. E. Max, and W. C. Mead, Theoretical interpretation of angle- and polarization-dependent laser light absorption measurements, *Phys. Fluids* **21**, 707 (1978).

9. Cairns, R. A., Resonant absorption at a rippled critical surface, *Plasma Phys.* **20**, 991 (1978).

10. David, F. and R. Pellat, Resonant absorption by a magnetic plasma at a rippled critical surface, *Phys. Fluids* **23**, 1682 (1980).

11. Manes, K. R., V. C. Rupert, J. M. Auerbach, P. Lee, and J. E. Swain, Polarization and angular dependence of $1.06\mu m$ light absorption by planar plasmas, *Phys. Rev. Lett.* **39**, 281 (1977).

12. Godwin, R. P., P. Sachsenmaier, and R. Sigel, Angle-dependent reflectance of laser-produced plasmas, *Phys. Rev. Lett.* **39**, 1198 (1977).

13. Pearlman, J. S. and M. K. Matzen, Angular dependence of polarization-related laser-plasma absorption processes, *Phys. Rev. Lett.* **39**, 140 (1977).

14. Balmer, J. E. and T. P. Donaldson, Resonance absorption of $1.06\mu m$ laser radiation in laser-generated plasma, *Phys. Rev. Lett.* **39**, 1084 (1977).

15. Villeneuve, D. M., G. D. Enright, and M. C. Richardson, Features of lateral energy transport in CO_2-laser-irradiated microdisk targets, *Phys. Rev. A* **27**, 2656 (1983).

16. Bach, D. R. *et al.*, Intensity-dependent absorption in 10.6μm laser illuminated spheres, *Phys. Rev. Lett.* **50**, 2082 (1983).

17. Burnett, N. H. and G. D. Enright, Hot electron generation and transport in high-intensity laser interaction, *Can. J. Phys.* **64**, 920 (1986).

18. Estabrook, K. G. and W. L. Kruer, Properties of resonantly heated electron distributions, *Phys. Rev. Lett.* **40**, 42 (1978).

19. Forslund, D. W., J. M. Kindel, and K. Lee, Theory of hot electron spectra at high laser intensity, *Phys. Rev. Lett.* **39**, 284 (1977).

20. Haas, R. *et al.*, Irradiation of parylene disks with a 1.06μm laser, *Phys. Fluids* **20**, 322 (1977).

21. Manes, K. R., H. G. Ahlstrom, R. H. Haas, and J. F. Holzrichter, Light-plasma interaction studies with high-power glass lasers, *J. Opt. Soc. Am.* **67**, 717 (1977).

22. Nuckolls, J., L. Wood, A. Thiessen, and G. Zimmerman, Laser compression of matter to super high densities: thermonuclear (CTR) applications, *Nature* **239**, 139 (1972).

23. Ripin, B. H. *et al.*, Enhanced backscatter with a structured laser pulse, *Phys. Rev. Lett.* **39**, 611 (1977).

24. Phillion, D. W., W. L. Kruer, and V. C. Rupert, Brillouin scatter in laser-produced plasmas, *Phys. Rev. Lett.* **39**, 1529 (1977).

25. Ng, A., L. Pitt, D. Salzmann, and A. A. Offenberger, Saturation of stimulated Brillouin backscatter in CO_2-laser-plasma interaction, *Phys. Rev. Lett.* **42**, 307 (1979).

26. Mayer, F. J., G. E. Busch, C. M. Kinzer, and K. G. Estabrook, Measurements of short-pulse backscatter from a gas target, *Phys. Rev. Lett.* **44**, 1498 (1980).

27. Rosen, M. D. *et al.*, The interaction of 1.06μm laser radiation with high Z disk targets, *Phys. Fluids* **22**, 2020 (1979).

28. Tanaka, K. *et al.*, Brillouin scattering, two-plasmon decay, and self-focusing in underdense ultraviolet laser-produced plasmas, *Phys. Fluids* **28**, 2910 (1985).

29. Turner, R. E. and L. M. Goldman, Evidence for multiple Brillouin modes in laser-plasma backscatter experiments, *Phys. Fluids* **24**, 184 (1981).

30. Herbst, M. J., C. E. Clayton, and F. F. Chen, Saturation of Brillouin backscatter, *Phys. Rev. Lett.* **43**, 1591 (1979).

31. Massey, R., K. Berggren and Z. Pietryzk, Observation of stimulated Brillouin backscattering from an underdense plasma, *Phys. Rev. Lett.* **36**, 963 (1976).

32. Goldman, L. M., W. Seka, K. Tanaka, R. Short, and A. Simon, The use of laser harmonic spectroscopy as a target diagnostic, *Can. J. Phys.* **64**, 969 (1986).

33. Gorbunov, L. M., Yu. S. Kasyanov, V. V. Korobkin, A. N. Polyanichev, and A. P. Shevelko, Spectral and temporal measurements of radiation of light backscattered by a laser plasma, *Sov. Phys. JETP* **27**, 226 (1978).

34. Maaswinkel, A., K.Eidmann, and R. Sigel, Comparative reflectance measurements of laser-produced plasmas at 1.06 and 0.53μm, *Phys. Rev. Lett.* **42**, 1625 (1979).

35. Nakai, S. *et al.*, Nonlinear interaction processes between a CO_2 laser and a plasma, *Phys. Rev.* **A17**, 1133 (1978).

36. Yamanaka, C., T. Yamanaka, T. Sasaki, J. Mizui, and H. B. Kang, Brillouin backscattering and parametric double resonance in laser-produced plasmas, *Phys. Rev. Lett.* **32**, 1038 (1974).

37. Basov, N. G. *et al.*, Generation of high power light pulses of wavelengths 1.06 and 0.53μm and their application in plasma heating, *Sov. J. Quant. Electron.* **2**, 439 (1973).

38. Mase, A. *et al.*, Stimulated-Brillouin-scattering studies in low density plasmas using microwave sources; in *Proceedings of the Eighth International Conference on Plasma Physics and Controlled Nuclear Fusion Research*, Vol. II, p.745. International Atomic Energy Agency, Vienna, 1981.

39. Pawley, C. J., N. C. Luhmann Jr., and W. L. Kruer, Microwave simulation of laser plasma intractions; in *Proceedings of the Eleventh International Conference on Plasma Physics and Controlled Nuclear Fusion Research*. International Atomic Energy Agency, Vienna, 1987.

40. Bobin, J. L., M. Decroisette, B. Meyer, and Y. Vitel, Harmonic generation and parametric excitation of waves in a laser-created plasma, *Phys. Rev. Lett.* **30**, 594 (1973).

41. Watt, R. G., R. D. Brooks, and Z. A. Pietryzk, Observation of stimulated Raman backscatter from a preformed, underdense plasma, *Phys. Rev. Lett.* **41**, 170 (1978).

42. Elazar, J., W. Toner, and E. R. Wooding, Backscattered radiation at $\omega_0/2$ from neodymium laser plasma interactions, *Plasma Phys.* **23**, 813 (1981).

43. Joshi, C., T. Tajima, J. M. Dawson, H. A. Baldis, and N. A. Ebrahim, Forward Raman instability and electron acceleration, *Phys. Rev. Lett.* **47**, 1285 (1981).

44. Figueroa, H., C. Joshi, H. Azechi, N. A. Ebrahim, and K. G. Estabrook, Stimulated Raman scattering, two-plasmon-decay and hot electron generation from underdense plasmas at .35μm, *Phys. Fluids* **27**, 1887 (1984).

45. Tanaka, K. *et al.*, Stimulated Raman scattering from UV-laser-produced plasmas, *Phys. Rev. Lett.* **48**, 1179 (1982).

46. Seka, W. *et al.*, Collective stimulated Raman scattering instability in UV laser plasmas, *Phys. Fluids* **27**, 2181 (1984).

47. Phillion, D. W., D. L. Banner, E. M. Campbell, R. E. Turner, and K. G. Estabrook, Stimulated Raman scattering in large plasmas, *Phys. Fluids* **25**, 1434 (1982).

48. Phillion, D. W., E. M. Campbell, K. G. Estabrook, G. E. Phillips, and F. Ze, High energy electron production by the Raman and $2\omega_{pe}$ instabilities in a 1.064μm laser-produced underdense plasma, *Phys. Rev. Lett.* **49**, 1405 (1982).

49. Turner, R. E. *et al.*, Evidence for collisional damping in high-energy Raman scattering experiments at .26μm, *Phys. Rev. Lett.* **54**, 189 (1985).

50. Turner, R. E. *et al.*, Observation of forward Raman scattering in laser-produced plasmas, *Phys. Rev. Lett.* **57**, 1725 (1986).

51. Offenberger, A., R. Fedosejevs, W. Tighe, and W. Rozmus, Stimulated Raman backscatter from a magnetically confined plasma column, *Phys. Rev. Lett.* **49**, 371 (1982).

52. Drake, R. P. *et al.*, Efficient Raman sidescatter and hot-electron production in laser-plasma interaction experiments, *Phys. Rev. Lett.* **53**, 1739 (1984).

53. Shephard, C. L. *et al.*, Raman scattering in experiments with planar Au targets irradiated with 0.53μm laser light, *Phys. Fluids* **29**, 583 (1986).

54. Kruer, W. L., Laser-driven instabilities in long scalength plasmas; in *Laser Plasma Interactions 3*, (M. B. Hooper, ed.) p.79–103. SUSSP Publications, Edinburgh, 1986.

55. Simon, A. and R. L. Short, New model of Raman spectra in laser-produced plasmas, *Phys. Rev. Lett.* **53**, 1912 (1984).

56. Simon, A., W. Seka, L. M. Goldman, and R. W. Short, Raman scattering in inhomogeneous laser-produced plasmas, *Phys. Fluids* **29**, 1704 (1986).

57. Baldis, H. A., J. C. Simpson, and P. B. Corkum, Two-plasmon decay and profile modification produced by 10.6μm radiation at quarter-critical density, *Phys. Rev. Lett.* **41**, 1719 (1978).

58. Baldis, H. A. and C. J. Walsh, Growth and saturation of the two-plasmon decay instability, *Phys. Fluids* **26**, 1364 (1983).

59. Meyer, J. and H. Houtman, Experimental investigation of the two-plasmon decay instability in a CO_2-laser-produced plasma, *Phys. Fluids* **28**, 1549 (1985).

60. Turner, R. E., D. W. Phillion, B. F. Lasinski, and E. M. Campbell, Half and three-halves harmonic measurements from laser-produced plasmas, *Phys. Fluids* **27**, 511 (1984).

61. Carter, P. D., S. M. L. Sim, H. C. Barr, and R. G. Evans, Time-resolved measurements of the three-halves harmonic spectrum from laser-produced plasmas, *Phys. Rev. Lett.* **44**, 1407 (1980).

62. Keck, R. L., R. L. McCrory, W. Seka, and J. M. Soures, Absorption physics at 351nm in spherical geometry, *Phys. Rev. Lett.* **54**, 1656 (1985).

63. Haas, R. A., M. J. Boyle, K. Manes, and J. E. Swain, Evidence of localized heating in CO_2-laser-produced plasmas, *J. Appl. Phys.* **47**, 1318 (1976).

64. Max, C. E. *et al.*, Scaling of laser-plasma interactions with laser wavelength and plasma size; in *Laser Interaction and Related Plasma Phenomena*, Vol. 6, p.507 (H. Hora and G. Miley, eds.). Plenum, New York, 1984.

65. Ng, A., D. Saltzmann, and A. A. Offenberger, Filamentation of CO_2-laser radiation in an underdense hydrogen plasma, *Phys. Rev. Lett.* **43**, 1502 (1979).

66. Tanaka, K., L. M. Goldman, W. Seka, R. W. Short, and E. A. Williams, Spectroscopic study of scattered light at around the fundamental wavelength in UV laser-produced plasmas, *Phys. Fluids* **27**, 2960 (1984).

67. Willi, O., P. T. Rumsby, and Z. Lin, Filamentation instability in laser-produced plasmas; in *Laser Interaction and Related Plasma Phenomena*, Vol. 6, p.633 (H. Hora and G. Miley, eds.). Plenum, New York, 1984.

68. Herbst, M. J., J. A. Stamper, R. R. Whitlock, R. H. Lehmberg, and B. H. Ripin, Evidence from x-ray, $\frac{3}{2}\omega_0$, and $2\omega_0$ emission for laser filamentation in a plasma, *Phys. Rev. Lett.* **46**, 328 (1981).

69. Baldis, H. A. and P. B. Corkum, Self-focusing of $10.6\mu m$ radiation in an underdense plasma, *Phys. Rev. Lett.* **45**, 1260 (1980).

70. Basov, N. G. *et al.*, Compression of laser irradiated hollow microspheres; in *Plasma Physics: Nonlinear Theory and Experiments*, p.47 (H. Wilhelmsson, ed.). Plenum, New York, 1977.

71. Yamanaka, C., T. Yamanaka, T. Sasaki, J. Mizui, and H. B. Kang, Brillouin backscattering and parametric double resonance in a laser-produced plasma, *Phys. Rev. Lett.* **32**, 1038 (1974).

72. Tanaka, K. *et al.*, Evidence of parametric instabilities in second harmonic spectra from 1054nm laser-produced plasmas, *Phys. Fluids* **27**, 2187 (1984).

73. Walsh, C. J., H. A. Baldis, and R. G. Evans, Plasma fluctuations at critical density in a CO_2 laser plasma interaction, *Phys. Fluids* **25**, 2326 (1982).

74. Chu, T. K. and H. W. Hendel, Measurements of enhanced absorption of electromagnetic waves and effective collision frequency due to parametric decay instability, *Phys. Rev. Lett.* **29**, 634 (1972).

75. Dreicer, H., R. F. Ellis, and J. C. Ingraham, Hot electron production and anomalous microwave absorption near the plasma frequency, *Phys. Rev. Lett.* **31**, 426 (1973).

76. Porkolab, M., V. Arunasalam, N. C. Luhmann Jr., and J. Schmitt, Parametric instabilities and plasma heating in an inhomogeneous plasma, *Nucl. Fusion* **16**, 269 (1976).

77. Mizuno, K., J. S. DeGroot, and K. G. Estabrook, Electron heating caused by parametrically driven turbulence near the critical density, *Phys. Fluids* **29**, 568 (1986).

78. Fejer, J. A., Ionospheric modification and parametric instabilities, *Rev. Geophys. Space Phys.* **17**, 135 (1979).

79. Wong, A. Y., J. Santoru, C. Darrow, L. Wang and J. G. Roederer, *Radio Science* **18**, 815 (1983).

80. Stamper, J. A. and B. H. Ripin, Faraday-rotation measurements of megagauss magnetic fields in laser-produced plasmas, *Phys. Rev. Lett.* **34**, 138 (1975).

81. Raven, A. *et al.*, Dependence of spontaneous magnetic fields in laser produced plasmas on target size and structure, *Appl. Phys. Lett.* **35**, 526 (1979).

82. Max, C. E. and K. G. Estabrook, Wavelength scaling in laser fusion from a plasma physics viewpoint, *Comments on Plasma Physics and Controlled Fusion* **5**, 239 (1980).

83. Kruer, W. L., Laser plasma coupling in reactor-size targets, *Comments on Plasma Physics and Controlled Fusion* **6**, 167 (1981).

84. Garban-Labaune, C. *et al.*, Effect of laser wavelength and pulse duration on laser-light absorption and back reflection, *Phys. Rev. Lett.* **48**, 1018 (1982).

85. Seka, W. *et al.*, Measurements and interpretation of the absorption of .35μm laser radiation on planar targets, *Opt. Comm.* **40**, 437 (1982).

86. Slater, D. C. *et al.*, Absorption and hot-electron production for 1.05 and 0.53μm light on spherical targets, *Phys. Rev. Lett.* **46**, 1199 (1981).

87. Nishimura, H. *et al.*, Experimental study of wavelength dependences of laser-plasma-coupling, transport, and ablation processes, *Phys. Rev. A* **23**, 2011 (1981).

88. Mead, W. C. *et al.*, Laser irradiation of disk targets at 0.53μm wavelength, *Phys. Fluids* **26**, 2316 (1983).

89. Ripin, B. H. and W. L. Kruer, Inertial confinement fusion systems; in *Plasma and Fluids—Physics Through the 1990s*, p. 221–236. National Academy Press, Washington, 1986.

90. Campbell, E. M., Dependence of laser-plasma interaction physics on laser wavelength and plasma scalelength; in *Radiation in Plasmas*, Vol. 1, p. 579 (B. McNamara, ed.) and World Scientific, Singapore, 1984.

91. Lehmberg, R. H. and S. P. Obenschain, Use of induced spatial inchoherence for uniform illumination of laser fusion targets, *Opt. Comm.* **46**, 27 (1983).

92. Kato, Y. *et al.*, Random phasing of high-power lasers for uniform target acceleration and plasma instability suppression, *Phys. Rev. Lett.* **53**, 1057 (1984).

93. Ximing Deng, Xiangchun Liang, Zezun Chen, Wenyan Yu, and Renyong Ma, Uniform illumination of large focal targets using a lens array, *Chin. J. Lasers* **12**, 257 (1985).

94. Obenschain, S. P. et al., Laser-target interaction with induced spatial incoherence, *Phys, Rev. Lett.* **56**, 2807 (1986).

95. Baldis, H. A., E. M. Campbell, and W. L. Kruer, in *Physics of Laser Plasmas* (A. Rubenchik and S. Witkowski, eds.), pp. 361–434, North-Holland, Amsterdam, 1991.

96. Kruer, W. L., Intense laser plasma interactions: from Janus to Nova, *Phys. Fluids* **B3**, 2356 (1991).

97. Young, P.E., J. H. Hammer, S. C. Wilks, and W. L. Kruer, Laser beam propagation and channel formation in underdense plasmas, *Phys. Plasmas* **2**, 2825 (1995).

98. Moody, J. D. *et al.*, First observation of intensity dependent laser beam deflection in a flowing plasma, *Phys. Rev. Lett.* **77**, 1294 (1996).

99. Kirkwood, R. K. *et al.*, Observation of energy transfer between frequency mismatched laser beams in a large scale plasma, *Phys. Rev. Lett.* **76**, 2065 (1996)

100. Fernandez, J. C. *et al.*, Observed dependence of stimulated Raman scattering on ion-acoustic wave damping in hohlraum plasmas, *Phys. Rev. Lett.* **77**, 2702 (1996); Kirkwood, R. K. *et al.*, The effect of ion wave damping on stimulated Raman scattering in high Z laser produced plasmas, *Phys. Rev. Lett.* **77**, 2706 (1996).

101. Montgomery, D. S. *et al.*, Evidence of plasma fluctuations and their effect on the growth of stimulated Brillouin and stimulated Raman scattering in laser plasmas, *Phys. Plasmas* **5**, 1973 (1998).

102. Labaune, C. *et al.*, Time-resolved measurements of secondary Langmuir waves produced by the Langmuir decay instability in a laser-produced plasma, *Phys. Plasmas* **5**, 234 (1998).

103. Montgomery, D. S., *et al.*, Recent Trident single hot spot experiments: evidence for kinetic effects and observation of Langmuir decay instability cascade, *Phys. Plasmas* **9**, 2311 (2002)

104. MacGowan, B. J. *et al.*, Laser plasma interactions in ignition-scale hohlraum plasmas, *Phys. Plasmas* **3**, 2029 (1996)
105. Glenzer, S. H. *et al.*, Energetics of inertial confinement fusion hohlraum plasmas, *Phys. Rev. Lett.* **80**, 2845 (1998)
106. Kauffman, R. L. *et al.*, Improved gas-filled hohlraum performance on Nova with beam smoothing, Phys. Plasmas 5, 1927 (1998).
107. Lindl, J. D., Inertial Confinement Fusion, Springer, New York, 1998.
108. Kruer, W. L. , Interaction of plasmas with intense lasers, *Phys. Plasmas* **7**, 2270 (2000)
109. Key, M. H. *et al.*, Hot electron production and heating by hot electrons in fast ignitor research, *Phys. Plasmas* **5**, 1960 (1998).
110. Wharton, K. *et al.*, Experimental measurements of hot electrons generated by ultrai-intense laser plasma interactions on solid-density targets, *Phys. Rev. Lett.* **81**, 822 (1998).
111. Kodama, R. *et al.*, Fast heating of ultrahigh-density plasma as a step towards fast ignition, *Nature* **412**, 798 (2001).
112. Clayton, C. E., *et al.*, Acceleration and scattering of injected electrons in plasma beat wave accelerator experiments, *Phys. Plasmas* **1**, 1753 (1994)
113. Esarey, E. P. , P. Sprangle, J. Krall, and A. Ting, Overview of plasma-based accelerator concepts, IEEE Trans. *Plasma Sci.*, **PS–24**, 252 (1996)
114. Umstadter, D., Review of physics and applications of relativistic plasmas driven by ultra-intense lasers, *Phys. Plasma* **8**, 1774 (2001)

Index

Printed in the United States
by Baker & Taylor Publisher Services